# DYNAMIC CHANGES
## IN MARINE ECOSYSTEMS
### Fishing, Food Webs, and Future Options

Committee on Ecosystem Effects of Fishing: Phase II—
Assessments of the Extent of Change and the Implications for Policy

Ocean Studies Board

Division on Earth and Life Studies

NATIONAL RESEARCH COUNCIL
*OF THE NATIONAL ACADEMIES*

THE NATIONAL ACADEMIES PRESS
Washington, D.C.
**www.nap.edu**

THE NATIONAL ACADEMIES PRESS 500 Fifth Street, N.W. Washington, DC 20001

NOTICE: The project that is the subject of this report was approved by the Governing Board of the National Research Council, whose members are drawn from the councils of the National Academy of Sciences, the National Academy of Engineering, and the Institute of Medicine. The members of the committee responsible for the report were chosen for their special competences and with regard for appropriate balance.

This study was supported by Contract/Grant No. DG133R04CQ0009 between the National Academy of Sciences and the National Oceanic and Atmospheric Administration. Any opinions, findings, conclusions, or recommendations expressed in this publication are those of the author(s) and do not necessarily reflect the views of the organizations or agencies that provided support for the project.

This report is funded in part by a contract from the National Oceanic and Atmospheric Administration. The views expressed herein are those of the author(s) and do not necessarily reflect the views of NOAA or any of its subagencies.

International Standard Book Number 0-309-10050-X (Book)
International Standard Book Number 0-309-65475-0 (PDF)

Library of Congress Catalog Number 2006927390

Cover art by Ray Troll, "North Pacific Marine Life," © 1986

Additional copies of this report are available from the National Academies Press, 500 Fifth Street, N.W., Lockbox 285, Washington, DC 20055; (800) 624-6242 or (202) 334-3313 (in the Washington metropolitan area); Internet, http://www.nap.edu.

Copyright 2006 by the National Academy of Sciences. All rights reserved.

Printed in the United States of America

# THE NATIONAL ACADEMIES
*Advisers to the Nation on Science, Engineering, and Medicine*

The **National Academy of Sciences** is a private, nonprofit, self-perpetuating society of distinguished scholars engaged in scientific and engineering research, dedicated to the furtherance of science and technology and to their use for the general welfare. Upon the authority of the charter granted to it by the Congress in 1863, the Academy has a mandate that requires it to advise the federal government on scientific and technical matters. Dr. Ralph J. Cicerone is president of the National Academy of Sciences.

The **National Academy of Engineering** was established in 1964, under the charter of the National Academy of Sciences, as a parallel organization of outstanding engineers. It is autonomous in its administration and in the selection of its members, sharing with the National Academy of Sciences the responsibility for advising the federal government. The National Academy of Engineering also sponsors engineering programs aimed at meeting national needs, encourages education and research, and recognizes the superior achievements of engineers. Dr. Wm. A. Wulf is president of the National Academy of Engineering.

The **Institute of Medicine** was established in 1970 by the National Academy of Sciences to secure the services of eminent members of appropriate professions in the examination of policy matters pertaining to the health of the public. The Institute acts under the responsibility given to the National Academy of Sciences by its congressional charter to be an adviser to the federal government and, upon its own initiative, to identify issues of medical care, research, and education. Dr. Harvey V. Fineberg is president of the Institute of Medicine.

The **National Research Council** was organized by the National Academy of Sciences in 1916 to associate the broad community of science and technology with the Academy's purposes of furthering knowledge and advising the federal government. Functioning in accordance with general policies determined by the Academy, the Council has become the principal operating agency of both the National Academy of Sciences and the National Academy of Engineering in providing services to the government, the public, and the scientific and engineering communities. The Council is administered jointly by both Academies and the Institute of Medicine. Dr. Ralph J. Cicerone and Dr. Wm. A. Wulf are chair and vice chair, respectively, of the National Research Council.

**www.national-academies.org**

# COMMITTEE ON ECOSYSTEM EFFECTS OF FISHING: PHASE II—ASSESSMENTS OF THE EXTENT OF CHANGE AND THE IMPLICATIONS FOR POLICY[1]

**JOHN J. MAGNUSON** (*Chair*), University of Wisconsin, Madison
**JAMES H. COWAN, JR.**, Louisiana State University, Baton Rouge
**LARRY B. CROWDER**, Duke University, Beaufort, North Carolina
**DORINDA G. DALLMEYER**, University of Georgia, Athens
**RICHARD B. DERISO**, Inter-American Tropical Tuna Commission, La Jolla, California
**ROBERT T. PAINE**, University of Washington, Seattle
**ANA M. PARMA**, Centro Nacional Patagónico, Chubut, Argentina
**ANDREW A. ROSENBERG**, University of New Hampshire, Durham
**JAMES E. WILEN**, University of California, Davis

**Staff**

**CHRISTINE BLACKBURN**, Program Officer
**SUSAN PARK**, Associate Program Officer
**NANCY CAPUTO**, Research Associate
**PHILLIP LONG**, Program Assistant

The work of this committee was overseen by the Ocean Studies Board.

---

[1] The committee and staff biographies are provided in Appendix A.

## OCEAN STUDIES BOARD

**SHIRLEY A. POMPONI** (*Chair*), Harbor Branch Oceanographic Institution, Fort Pierce, Florida
**LEE G. ANDERSON**, University of Delaware, Newark
**JOHN A. ARMSTRONG**, IBM Corporation (retired), Amherst, Massachusetts
**WHITLOW AU**, University of Hawaii at Manoa
**ROBERT G. BEA**, University of California, Berkeley
**ROBERT DUCE**, Texas A&M University, College Station
**MARY (MISSY) H. FEELEY**, ExxonMobil Exploration Company, Houston, Texas
**HOLLY GREENING**, Tampa Bay National Estuary Program, St. Petersburg, Florida
**DEBRA HERNANDEZ**, Hernandez and Company, Isle of Palms, South Carolina
**CYNTHIA M. JONES**, Old Dominion University, Norfolk, Virginia
**WILLIAM A. KUPERMAN**, Scripps Institution of Oceanography, La Jolla, California
**FRANK E. MULLER-KARGER**, University of South Florida, St. Petersburg
**JOAN OLTMAN-SHAY**, NorthWest Research Associates, Inc., Bellevue, Washington
**ROBERT T. PAINE**, University of Washington, Seattle
**S. GEORGE H. PHILANDER**, Princeton University, New Jersey
**RAYMOND W. SCHMITT**, Woods Hole Oceanographic Institution, Massachusetts
**DANIEL SUMAN**, Rosenstiel School of Marine and Atmospheric Science, University of Miami, Florida
**STEVEN TOMASZESKI**, Rear Admiral, U.S. Navy (retired), Fairfax, Virginia
**ANNE M. TREHU**, Oregon State University, Corvallis

**Staff**

**SUSAN ROBERTS**, Director
**DAN WALKER**, Scholar
**FRANK HALL**, Program Officer
**SUSAN PARK**, Associate Program Officer
**ANDREAS SOHRE**, Financial Associate
**SHIREL SMITH**, Administrative Coordinator
**JODI BOSTROM**, Research Associate
**NANCY CAPUTO**, Research Associate
**SARAH CAPOTE**, Senior Program Assistant

# Preface

Challenges to sustaining the productivity of oceanic and coastal fisheries have become more critical and complex as these fisheries reach the upper limits to ocean harvests. In addition, it is now clear that we are managing interactive and dynamic food webs rather than sets of independent single-species populations. Fisheries products cannot be extracted from the sea without ecosystem effects; even though we all know this, we have not incorporated the consequences of fishing food webs and modifying trophic structure and species interactions into the scientific advice that informs policy and management systems. This insufficiency has come at a cost of collapsed fisheries and unintended consequences. Fisheries influence non-targeted as well as targeted species. Some of the non-targeted species are part of the bycatch, but others have been affected profoundly by the complex interactions in food webs initiated by fisheries that reduce the abundance of their predators or prey.

Publicity accompanying the publication of several prominent articles in the scientific literature on the influence of fisheries on apex predatory fishes and on the changing structure of marine food webs generated public concern that the oceans had been "fished out" quite literally. Our committee was charged with the review and evaluation of the current literature (including these high visibility papers) on the impacts of modern fisheries on the composition and productivity of marine ecosystems. After discussions about this assignment with the sponsor at our first committee meeting, it became clear that neither the committee nor the sponsor wanted a detailed peer review or a reanalysis of those scientific reports that attracted so much public attention. Instead, we determined that this study should provide an overview of the topic, including a review of these highly visible papers in the context of the broader body of literature now available. The report provides an overview of the influence of fisheries on marine food webs and productivity. We were also asked to discuss the relevance of these findings for U.S. fisheries management and to identify areas for future research and analysis. Lastly, we were asked to characterize the stewardship implications of our findings for living marine resources. This report and its findings will challenge scientists and managers to implement new approaches to fisheries policy and management.

The committee recognized from the onset that ecosystem effects on fishery productivity include other issues related to water quality and pollution, habitat modifications and loss, land use, invasive species, climatic change, and other factors. These need to be incorporated into an ecosystem-based approach to managing oceans and coasts. Such concerns were not in our charge, and we did not deal with them here. However, these drivers do impact fisheries dynamics and are as important to sustaining fishery productivity as those we do address.

We believe that moving from a single-species approach toward a food-web management approach is an important step forward in achieving an ecosystem approach to fisheries management. In this new context for fisheries management, scientists will be challenged to provide policy-relevant options; managers will be challenged to broaden their concerns and experiment openly; and policy makers will be challenged to act unselfishly on behalf of the broader community of people who value and depend on ocean ecosystems.

As the committee addressed its charge—to review and evaluate the impacts of modern fisheries on the composition and productivity of marine ecosystems and their relevance to U.S. fisheries management, future research and stewardship of living marine resources—certain overarching principles and concepts emerged repeatedly. Taking a long-term and broad spatial view at multiple scales of resolution and extent is essential. Synthesis and food-web modeling provide alternative scenarios that can more robustly inform harvest strategies than can analyses of single populations. Social sciences and the tradeoffs between different fisheries and fishermen infuse all decisions on how best to harvest different components of food webs and to allocate these ocean resources among users. Sustaining ecosystem services from the ocean is equally as important as managing consumptive uses such as fisheries. Unfortunately, non-consumptive uses and ecosystem services are poorly accounted for and represented in fishery research, policy, and management. We have a vision of how to incorporate food-web considerations into fisheries management, but we do not have a practice or a handbook; iterative examination and response to changes in fish populations and communities will be the rule if we are to better steer marine ecosystems using fishery policies.

The committee of nine included three fishery scientists, four aquatic ecologists, and two social scientists with broad knowledge of the issues. More specific information on the issues was presented by a broad group of scientists at the three meetings of the committee. We greatly appreciated their contributions to our deliberations.

I thank the committee members for their many contributions of text, ideas, and knowledge and their willingness to review, debate, and reach consensus. All members contributed and brought new information and insight to the process and valued judgment to the table. I thank and congratulate Dr. Christine Blackburn, our study director, who met the challenge of her first study committee at the National Research Council. I have been most pleased to work with her. I especially appreciate her dedication to the purpose of our task, her tireless effort to complete the report, her ability to learn, her demand for accuracy of the presented information, and her unselfish openness to debate and deliberation in order to reach consensus and synthesis. I thank Ms. Nancy Caputo, Research Associate, who has been a resourceful team member and whose imprint has greatly improved our report both broadly and in detail. I thank Mr. Phillip Long, Program Assistant, for facilitating our committee, our travels, and our teleconferences. These three are a good group.

John J. Magnuson, *Chair*

# Acknowledgments

This report was greatly enhanced by the participants of the three workshops held as part of this study. The committee would first like to acknowledge the efforts of those who gave presentations at meetings: Villy Christensen, University of British Columbia; Jeremy Collie, University of Rhode Island; Joshua Eagle, University of South Carolina; Timothy Essington, University of Washington; David Fluharty, University of Washington; Michael Fogarty, Northeast Fisheries Science Center, National Oceanic and Atmospheric Administration; Anne Hollowed, Alaska Fisheries Science Center; James Kitchell, University of Wisconsin; Phillip Levin, National Oceanic and Atmospheric Administration; Steven Murawski, National Oceanic and Atmospheric Administration; Daniel Pauly, University of British Columbia; Alison Rieser, University of Maine; Michael Sissenwine, National Oceanic and Atmospheric Administration; and William Sydeman, Point Reyes Bird Observatory. These talks helped set the stage for fruitful discussions in the closed sessions that followed.

This report has been reviewed in draft form by individuals chosen for their diverse perspectives and technical expertise, in accordance with procedures approved by the National Research Council's Report Review Committee. The purpose of this independent review is to provide candid and critical comments that will assist the institution in making its published report as sound as possible and to ensure that the report meets institutional standards for objectivity, evidence, and responsiveness to the study charge. The review comments and draft manuscript remain confidential to protect the integrity of the deliberative process. We wish to thank the following individuals for their participation in their review of this report:

**JEREMY S. COLLIE**, University of Rhode Island, Narragansett
**SERGE GARCIA**, U.N. Food and Agriculture Organisation (FAO), Rome, Italy
**RAY W. HILBORN**, University of Washington, Seattle
**JEREMY B. JACKSON**, University of California, San Diego, La Jolla
**MICHAEL K. ORBACH**, Duke University, Beaufort, North Carolina
**PIETRO PARRAVANO**, Commercial Fisherman, Half Moon Bay, California
**CLARENCE G. PAUTZKE**, North Pacific Research Board, Anchorage, Alaska
**VICTOR RESTREPO**, International Commission for the Conservation of Atlantic Tunas, Madrid, Spain
**CARL J. WALTERS**, University of British Columbia, Vancouver, Canada
**JAMES WILSON**, University of Maine, Orono

Although the reviewers listed above have provided many constructive comments and suggestions, they were not asked to endorse the conclusions or recommendations nor did they see the final draft of the report before its release. The review of this report was overseen by **John E. Burris**, Beloit College, Beloit, Wisconsin, and **May R. Berenbaum**, University of Illinois, Urbana, who were appointed by the National Research Council, and who were responsible for making certain that an independent examination of this report was carried out in accordance with institutional procedures and that all review comments were carefully considered. Responsibility for the final content of this report rests entirely with the authoring committee and the institution.

# Contents

SUMMARY                                                                    1

1   INTRODUCTION                                                          13
    Policy Context, 15
    Scientific Context, 17
    Policy Choices and the Role of Science, 17
    Moving Toward Ecosystem-Based Management, 19
    Committee Approach and Report Organization, 20

2   EVIDENCE FOR ECOSYSTEM EFFECTS OF FISHING                             23
    Changes in Abundance and Biomass, 24
    Genetic Changes in Populations, 32
    The Phenomena of Shifting Baselines, 33
    Altered Food Webs, 35
    Trophic Cascades, 41
    Fishing Down and Through the Food Web, 45
    Responding to Regime Shifts, 51
    Recovery, Stability, and Multiple Stable States, 53
    Major Findings and Conclusions for Chapter 2, 56

3   CONSIDERING THE MANAGEMENT IMPLICATIONS                               59
    Fisheries Management Implications of Ecosystem Interactions, 60
    Management Implications Aside from Trophic Interactions and
        Tradeoffs, 63

Developing Multiple Stock Harvest Strategies, 66
Mechanisms for Implementing Multi-Species Harvesting Strategies, 69
Overcoming Regulatory Constraints to Setting Multi-Species
   Reference Points, 74
Major Findings and Conclusions for Chapter 3, 75

4  INFORMING THE DEBATE: EXAMINING OPTIONS FOR          77
   MANAGEMENT AND STEWARDSHIP
   Evaluating Strategic Management Options, 78
   Projecting Recovery Strategies and the Effects of Shifting Baselines, 83
   Strategies for Informed and Inclusive Decision Making, 84
   Major Findings and Conclusions for Chapter 4, 90

5  SCIENCE TO ENABLE FUTURE MANAGEMENT                  93
   Improving Ecosystem Models and Scenario Analysis, 94
   Analyzing Historical Time-Series Data, 99
   Contributions from Social and Economic Science, 101
   Major Findings and Conclusions for Chapter 5, 105

6  FINDINGS AND RECOMMENDATIONS                        109
   Recommendations, 110

REFERENCES                                             119

APPENDIXES
A  Committee and Staff Biographies                     133
B  List of Acronyms                                    139
C  Committee Meeting Agendas                           141
D  Glossary                                            147

# Summary

Recent scientific literature has raised many concerns about whether fisheries have caused more extensive changes to marine populations and ecosystems than previously realized or predicted. Due to its extractive nature, fishing reduces stocks of harvested species. However, in many cases, stocks have been exploited far beyond management targets, ultimately reducing the potential productivity of the fishery. In addition, new analyses indicate that the abundance and composition of non-targeted organisms in marine ecosystems are radically changing as a result of fishing pressure expressed through food-web interactions.

Several scientific papers suggest that populations of high-trophic-level fishes have been severely depleted and that fishing has fundamentally altered the structure of marine ecosystems in many locations. But the conclusions drawn in these scientific papers often have been controversial. Subsequent articles have disputed the findings of these papers, and others have disputed the implications (or the broad application) of the conclusions presented, while still others continue to provide additional analyses. Arguments on all sides acknowledge the paucity of fishery-independent data as a major roadblock to properly analyzing the current state of fisheries and ecosystems. Instead, the analyses rely on the more readily available landing and catch statistics. These fishery-dependent data are subject to various interpretations because fisheries landings change in response to many factors other than the abundance of the fished stocks (e.g., markets, management regulations, fishing methods, technology, and climate).

While the fisheries science community continues to analyze and debate these issues, several of these publications have been widely publicized. This has increased public awareness and raised concern that fisheries resources are not being effectively managed, including the impacts of fishing on non-target resources and habitat. In response to this growing concern, the National Oceanic and Atmospheric Administration asked The National Academies' Ocean Studies Board to form a committee of experts to review recent scientific reports and weigh the collective evidence for fisheries-induced changes to the dynamics of marine ecosystems. The committee was asked to discuss the relevance of these scientific findings for U.S. fisheries management, to identify areas for future research and analysis, and to characterize the stewardship implications for living marine resources. To help accomplish these tasks, the committee met publicly three times to hear presentations on relevant subjects ranging from fisheries biology and fisheries governance mechanisms to current modeling and analysis techniques, among others.

## EVIDENCE FOR ECOSYSTEM CHANGE

Fishing can alter a wide range of biological interactions, causing changes in predator-prey relationships, cascading effects mediated through food-web interactions, and the loss or degradation of essential habitats. These impacts, along with natural fluctuations in the physical state of the ocean, can interact to intensify fishing impacts beyond targeted species. Fishing is also generally size and species selective, potentially changing the genetic structure and age composition of fished stocks, as well as decreasing the diversity of marine communities. Examples of all these effects have been documented. Although some changes are expected outcomes of management actions, in many instances the measured effects are quantitatively and qualitatively more severe than anticipated by management.

Declines in stock abundance have been measured for many species throughout the world's oceans, but the extent and severity of these declines differ across stocks and geographical areas. Changes to food-web interactions are expected because fisheries reduce the abundance of one or more components of the food web, simultaneously altering the interactions among species and the strength of these interactions. Direct predator-prey relationships have changed—either releasing lower trophic levels from predation or reducing the availability of prey for higher-level predators—and these effects may spread to successive trophic levels up and down the food web. Such cascading effects are often unforeseen and management actions frequently have unexpected results, especially if the target species plays a critical role in the ecosystem. Some of the greatest long-term impacts of fishing have been observed in non-targeted ecosystem components. Many species, including marine mammals, seabirds, sea turtles, sharks, oysters, kelps, and sea grasses, have been negatively affected by fisheries either directly through bycatch or habitat damage, or indirectly through altered food-web interactions.

One area of active inquiry is the underlying cause for the measured reduction in mean trophic level of landings seen in many of the world's oceans. Originally, these reductions were attributed to sequentially fishing lower trophic levels as higher ones were depleted, a process termed "fishing down the food web." A more recent analysis has offered an alternative hypothesis of "fishing *through* the food web," where fisheries add lower trophic species while continuing to catch higher trophic level species. These differing conclusions underscore two of the important issues addressed in this report. The first is the need for new models and new data to identify the underlying cause of change in marine ecosystems. The second is the recognition that the implications for management will differ based on this underlying cause. Fishing down the food web indicates that lower-trophic-level-species are harvested due to the depletion of the higher level predators. Fishing through the food web indicates that multiple trophic levels are being fished simultaneously—and perhaps sustainably. The appropriate management action for each of these cases should be crafted based on the specifics of the ecosystem, species, fishing methods, and values involved.

Whether the unwanted, negative influences of fishing on marine food webs and communities can be reversed is generally unknown. While some stocks have experienced recovery when fishing pressure was reduced, others have not. The overall productivity and composition of marine ecosystems may change for systems exploited beyond a certain threshold with no guarantee of reversibility—  new states may persist and even resist return to earlier conditions. In addition, environmental changes, such as climate-driven regime shifts, affect fishery productivity, creating conditions where recovery is even more uncertain.

## TRADEOFFS IN MANAGING MARINE FISHERIES

Management decisions for a particular targeted stock will have impacts on the productivity of other interacting species. Accounting for species linkages in a management context requires that harvest strategies for each species be chosen in ways that recognize the interconnectedness of marine ecosystems. In addition, other consumptive uses, nonconsumptive uses (e.g., recreation and scenic opportunities), and ecosystem services (e.g., nutrient cycling and climate and weather regulation) should be considered when formulating ecosystem goals. Because it is unlikely that value and yield can be simultaneously maximized for all services, tradeoffs are inevitable among various uses and services provided by the marine environment.

Scientific knowledge, from both natural and social sciences, is important for delineating options and illuminating choices, but allocation tradeoffs are public policy decisions. Various stakeholder groups will value a different mix of resource uses and desire different outcomes from management activities. Decisions about what mix of services the ocean will provide and what protections will be afforded to ocean species should be made with input from a broad range of stakeholders. Ultimately, a flexible management structure is needed to adapt to shifting ecosystem dynamics and changing stakeholder values, as well as to integrate decision making across all sectors of human activity. If decisions about tradeoffs in ecosystem services are to be equitable, fisheries management decisions will require consideration of other nonfishery uses of the marine environment.

## KEY FINDINGS AND RECOMMENDATIONS

Ecosystem-level effects of fishing are well supported in the scientific literature, including changes in food-web interactions and fluctuations in ecosystem productivity. Stock biomass and abundance have been reduced by fishing, and the size structure of populations has been altered. Furthermore, changes in trophic structure, species interactions, and biodiversity have been discovered, and fisheries-induced alternative ecosystem states (defined by a different species composition or productivity than that of the prefishing condition) are possible. Assuming that the upper level of harvest productivity from wild ocean resources is at or approaching the theoretical limit, and recognizing the inability to change one ecosystem component without affecting numerous others, food-web interactions will become increasingly important in future fisheries management decisions. Society will need to determine which ecosystem components are the most desirable for harvest, and then managers will need to implement policies designed to maximize this desired production while recognizing that this will affect other species.

If the United States is to manage fisheries within an ecosystem context, food-web interactions, life-history strategies, and trophic effects will need to be explicitly accounted for when developing harvesting strategies. Other uses and values derived from marine resources should also be considered, because fishing activities directly or indirectly impact other ecosystem components and the goods and services they provide. A modeling framework is necessary to examine ecosystem interactions and to compare the possible outcomes of different fishery management actions. Decisions about management strategies should be made in a manner that accounts for the range of uses involved and their relative social, ecological, and economic values.

### Applying Scenario-Based Decision Making

Currently, fisheries management approaches in the United States do not explicitly account for the ecosystem-level impacts discussed in this report. Furthermore, existing policies do not generally consider the possible effects of fisheries on other services provided by the ocean environment.

**Multiple-species harvest strategies should be evaluated to account for species interactions and food-web dynamics.**

Setting multi-species harvest strategies requires taking into account food-web interactions, changes in trophic structure and species interactions, life-history strategies, and bycatch. If management is to account for the ecological interdependence among harvest targets and other food-web components, it will be necessary to quantitatively and qualitatively examine these interactions. Increased application of food-web, species-interaction, and ecosystem models, and development of new models could provide a better understanding of food-web effects and the impacts of fishing on ecosystem components and help to develop multi-species harvest strategies.

**Food-web, species-interaction, and ecosystem models should be used to evaluate alternative policy and management scenarios. These scenarios should inform the choice of multi-species harvest strategies by elucidating the tradeoffs that will be required from the various user communities to manage in a multi-species context.**

Presently, fishery management policies employed in the Unites States are prescriptive, defined in terms of nonspecific biological reference points used to set target and limit harvest rates and to specify biomass thresholds for single species. The basic stock assessment process largely informs tactical decisions, rather than evaluating the consequences of different policy choices for all stakeholders. However, in an ecosystem context, management decisions will reflect value judgments and tradeoffs between uses; hence, scientific advice should provide strategic options about different management scenarios that can then be debated in the public-policy arena.

Ecosystem and food-web models exist that can provide useful tools for evaluating policy alternatives. The challenge for scientists and managers is to identify and assign probabilities to a range of scenarios that capture existing uncertainties about food-web dynamics and responses of food webs to various fishing strategies. The proposed approach includes the creation of appropriate model scenarios for managed systems, the generation of a number of management strategies to be evaluated, and the determination of performance statistics for measuring policy outcomes that will reflect the interests of all stakeholders. The alternative scenarios may represent different structural models for the dynamics and current status of the interactive system of species, different levels of productivity, and maximum population sizes under various climate regimes, as well as different relationships between the performance indicators.

The creation of integrated biologic-socioeconomic models will help to make tradeoff decisions even more explicit and informative. Ideally, new models will be able to capture important biophysical linkages and human impacts via economic market valuation methods. The most useful models will be those that include not only the best depictions of ecosystem links, but also accurate depictions of fishermen's behaviors and responsiveness to changes in governance systems.

Scenario analyses and the corresponding management actions are best applied in an iterative and adaptive process (Figure S.1). Monitoring programs should be

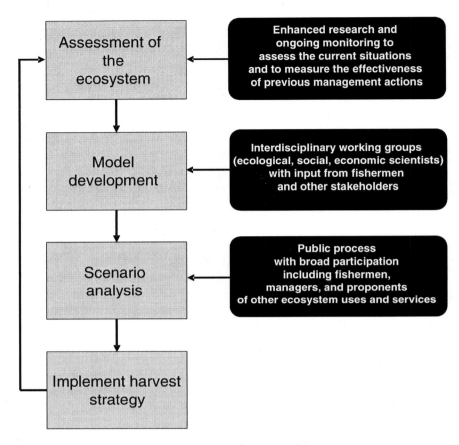

**FIGURE S.1** The process of scenario analysis-based management should be an iterative adaptive process. Improved data on food-web interactions, and changes in these interactions in both time and space, will help to create and update the models developed for a particular system. New and traditional regulatory schemes (catch and effort quotas set by different feedback control rules, marine protected areas, slot limits, gear type, etc.) and different monitoring schemes can, in principle, all be tested for their potential impacts on fished ecosystems and on user groups through the analysis process. Further, it is desirable that future models be set up to analyze the outcomes of different economic and social dynamics, behavior, and market pressures. Once there is a way to visualize all these different options, then a broad range of stakeholders can discuss which management schemes best achieve their collective goals and what tradeoffs are involved in deciding the management action that should be taken. Monitoring and regular assessments will be needed to feed the management process and to determine how well the previous actions achieve the intended outcome, and data should be collected on how essential ecosystem components changed. This information will then feed back into model development, and a new round of evaluating alternative management strategies will be initiated.

an integral component of management. Data are necessary to evaluate how both marine organisms and fishermen respond to changing management actions. Models will improve as more is learned and greater levels of complexity are added, requiring an adaptive approach to management.

**Interdisciplinary working groups should be considered as a mechanism for developing appropriate models for each management area and for generating the series of scenarios needed to test proposed management actions.**

Building models relevant for fisheries management will require the cooperation of many specialists and the integration of information from many sources. Working groups can provide a mechanism for bringing together scientists from government and academia as well as natural and social sciences in order to examine particular areas or fisheries of concern. Including social and economic scientists at the beginning of this process will ensure that these issues are incorporated as the base model is created.

Such working groups would facilitate consolidation of existing information, generate new syntheses with existing models, and develop new models and other new approaches to inform scenario development and forecasting under alternative management strategies. Working groups could meet with a variety of stakeholders—including fishermen and other consumptive and nonconsumptive users—to identify important tradeoffs that should be considered when creating models and evaluating feasible candidate policies. The simulations created should be quantitative when possible, but even rigorous qualitative scenarios would be useful in some situations. Iterative analyses by the working group might be expected as the ecosystem responds to management actions.

### Implementing Mechanisms for Multi-species Management

Fisheries are primarily managed by direct or indirect controls on either inputs (e.g., effort, gear type and configuration, time, and area closures or openings) or outputs (e.g., catch in weight or numbers; limitations on landing according to size, sex, or species; and maximum bycatch amounts). Most fisheries management in the United States and internationally relies on output controls with catch quotas as a primary regulatory objective, accomplished by some input controls on gear, areas, and seasons. From an ecosystem perspective, addressing the manner in which fishing is conducted via input controls may be more important than limiting the outputs. This is because ecosystem effects often result from the specifics of the fishing methods, rather than the absolute level of target-species removal.

Two approaches exist for regulating fishing effort to achieve either single-species or multi-species management objectives. By far, the most common method used in both the United States and internationally is top-down control. In this report, top-down control refers to a system in which harvest targets or limits are set by a management body, often with stakeholder participation, and then input controls are chosen and implemented to achieve these goals. Alternatively, bottom-up management approaches that confer secure access privileges are available. These types of management instruments still require that harvest targets be set by a centralized management body, but the details of effort and input decisions are decentralized.

In principle, existing top-down regulatory procedures can be adapted to account for ecosystem-level effects of fishing. However, a potential benefit of secure access privileges is that they can foster a stewardship ethic among fishermen motivated by concern for the long-term health and productivity of the fishery. These approaches may also promote fishing innovations that reduce impacts to other ecosystem components if access to the fishery is predicated on limiting impacts on non-target species.

**New governance and management instruments that create stewardship incentives among user groups should be evaluated and considered for adoption in the United States for multi-species management.**

Individual quotas, harvester cooperatives, community cooperatives, and territorially defined cooperatives exist in a handful of fisheries in the United States and in other countries. However, new research is needed to understand how these systems affect incentives in a multi-species setting, and how they might be adapted to handle more inclusive ecosystem goals associated with fisheries management in the United States.

### Incorporating Additional Values in Fisheries Management

Consumptive uses are those that rely on the removal or harvest of ocean resources, such as fishing, and therefore their value is readily measured based on market demand. On the other hand, nonconsumptive values and the value of ecosystem services are much more difficult to measure. The most common nonconsumptive values are those related to tourism, research, and education, in which values are expected to positively correlate with healthy ecosystems. The value provided by ecosystem services such as nutrient cycling and weather regulation are extremely difficult to quantify, but may be proportionally more important—ecosystem services are experienced across society, although the values are often overlooked. In order to make informed decisions about the suite of services provided by ocean ecosystems, increased understanding is needed about the range of values generated by these systems and about how these values affect different stakeholder groups.

**Fisheries management structures should ensure that a broad spectrum of social values is included in policy and management decisions.**

A diverse cross-section of constituents may be needed to advise decision makers on the desired balance of ecosystem values and uses that management should try to achieve. An important public policy issue is how to ensure that nonconsumptive and public-good values receive proper consideration when making tradeoffs among ocean services. Further, incorporating a broader range of values will require input from fisheries, ecosystem, and social scientists to help understand how various ecosystem configurations generate services that are valued by different stakeholder groups. Melding fisheries science and social science will be important for understanding how behavior might be modified in response to changing priorities and management actions.

### Supporting Research

Implementing scenario-based analysis; considering alternative management instruments; and integrating ecological, social, and economic values into fisheries management decisions require enhanced research in a number of areas. Scientific advances will need to incorporate new ideas, analyses, models, and data; perhaps, more importantly, new social and institutional climates will need to be established that catalyze a creative, long-term, comparative, and synthetic science of food webs and communities. Data needed to support ecosystem-based management will likely be more than the simple sum of currently available single-species information. Where species interact and to what extent will be as important as determining a stock biomass. Furthermore, a rich array of social science, economic science, and policy considerations will be essential, because many more tradeoffs are likely to be apparent among ecosystem components and stakeholders.

**Research is needed to improve our understanding of the extent of fishing effects on marine ecosystems and to promote the development of ecosystem, food-web, and species-interaction models for incorporation into management decisions.**

Promising results have come from analyses and models at levels of synthesis above individual populations and individual food-web components. However, if these models are to be applied in a management setting, greater knowledge of trophic effects and species interactions is necessary. Modification of existing models and/or the development of new models are needed to better account for uncertainty in model output, to elucidate indicators of regime shifts and other interacting factors, and to evaluate monitoring schemes necessary to provide adequate information on ecosystem structure and function.

Support of research in a number of areas will help to improve the utility of current and future models, including quantifying important food-web interactions, per capita and population interaction strengths, and baseline abundance data on a number of non-target and lower-trophic-level species. Much more can be learned about food-web linkages and interactions, including the strength of linkages between species and life-history stages and how these interactions change over time. Because so little is known about the complexities of marine ecosystems, data needs should be prioritized both for near- and long-term efforts, and for species and areas of concern.

Spatial analyses may be one of the greatest obstacles faced by fishery managers, yet new developments in measurement and analysis methods allow for the explicit consideration of spatial variability in marine systems. Collecting spatially explicit biological data will be essential for monitoring and assessing both large-scale population trends and changes at finer scales. Patterns of interaction and the strength of these interactions vary in time and space. Collecting data in both dimensions will increase understanding about the potential variability in these interactions and advance the ability of models to represent future scenarios. Furthermore, biologically relevant boundaries in the marine environment are virtually impossible to identify. Research is needed to determine whether ecosystem boundaries, for both modeling endeavors and management, could be better defined based on known interactions. In addition, analyzing population trends and species interactions on finer spatial scales may lead to new ideas about temporally and spatially flexible, area-based management.

Looking back in time is as important as assessing current ecosystem status. Assessments of historical data can provide new insights about former species abundances and interactions. A historical perspective is important for many reasons, chief among them is avoiding the shifting baselines phenomena. If recovery goals are to be established, it is wise to comprehend the levels of ecosystem productivity that were once possible. Further, synthesizing these types of data using models may allow for the examination of past interactions and their relative importance, indicating when it might be desirable to try to restore these interactions. Landings data, narratives and descriptions, fisheries-independent data, phytoplankton and plankton records, satellite data, archived specimens, empirical knowledge, and many other sources of information should all be considered when conducting these types of analyses.

**Research is needed to expand relevant social and economic information and to integrate this knowledge into fisheries management actions.**

Understanding social and ethical values linked to the broad suite of services provided by marine ecosystems is essential and will require measuring and scaling of those values in relation to other uses. While some valuation analyses have been completed for terrestrial systems, little comparable work exists for marine systems. Once it is hypothesized how various fishing strategies affect the structure and functioning of marine ecosystems, methods can be designed to evaluate how these changes affect humans directly and indirectly, elucidating those policy options that reflect the most desirable choices.

Evaluating management options will require integrated models that incorporate not only the best depictions of ecosystem links, but also the most accurate depictions of fishermen's behaviors and responsiveness to changes in regulations and governance systems. As mentioned previously, integrated biologic-socioeconomic models should be explored for their capacity to capture important biophysical linkages that are translated through human impacts via economic market valuation methods. Understanding behavior is a particularly under-researched area, even behavior associated with conventional management systems. Information should be collected to examine how different kinds of governance mechanisms could potentially change fishermen's behavior. Integrating this information into combined models would allow for the explicit consideration of all aspects of the management process—the how and where of biological resource interactions and the how, where, and why of human actions.

Finally, research should be conducted on how new governance mechanisms might better align fishing incentives to address more encompassing ecosystem management objectives. Most existing incentive-based systems are primarily single-species focused, but some are also beginning to address broader ecosystem objectives. Existing experiences with individual quotas, harvester cooperatives, community cooperatives, and territorially defined cooperatives should be examined. New research is also needed on management strategies that might best address food-web linkages, bycatch questions, and broader portfolios of ecosystem services. Analyzing available experiences worldwide can indicate whether these systems might be appropriate for adoption in U.S. fisheries to reduce ecosystem-level impacts of fishing.

# 1

# Introduction

Fishing by its very nature alters marine ecosystems by selectively removing fish and invertebrates. Humans have become one of the oceans' dominant predators, and with human populations continuing to grow, the influence on the marine environment is escalating. Today, a significant portion of energy and protein from fish and invertebrates is directed toward human uses (Watson and Pauly 2001, Food and Agriculture Organization [FAO] 2002) (see Appendix B for list of acronyms). However, the upper limit of potential harvest from wild ocean resources may have been reached (Garcia and Grainger 2005). Seventy-six percent of the world's stocks are fully exploited, overexploited, or depleted; few resources remain that may provide for the development of additional sustainable fisheries (Hilborn et al. 2003, FAO 2005) (Figure 1.1).

Concerns about overfishing have been expressed by fisheries specialists for decades, but in the last 10 to 15 years, overfishing has become a major public issue (Mace 2004). New analyses emphasize the multi-species nature of fisheries and indicate that the resulting changes in predation and competition should be accounted for in fishery management approaches (May 1984, International Council for the Exploration of the Sea [ICES] 2000, Sinclair and Valdimarsson 2003). The issue has not been whether this should be done, but continues to be how it can be done. A number of recent scientific publications conclude that changes to marine populations and food webs caused by fisheries removals are larger than had been previously believed, raising public and scientific concerns about the true extent of changes in marine ecosystems. At the same time, both the U.S. Commission on Ocean Policy (2004) and the Pew Oceans Commission (2003) recommend managing resources in an ecosystem context, including

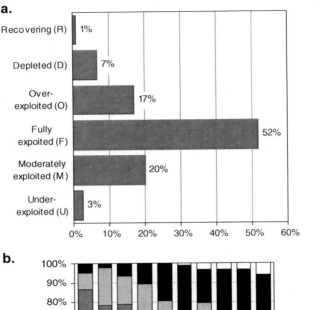
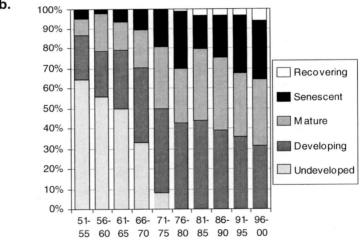

**FIGURE 1.1** (a) A 2004 global assessment of 441 stocks shows that over 75 percent are fully exploited, overexploited, or depleted. (b) Furthermore, there is a clear trend since the early 1950s in the top 200 global fisheries. The proportion of "undeveloped" resources fell to zero by the mid 1970s. The proportion of "developing" resources has decreased since the early 1990s. The "mature" resources have kept increasing since the beginning of the series. The fact that over two-thirds of these resources appear either mature, senescent, or recovering underscores the fact that we may be approaching global capacity for fisheries productivity. (Undeveloped: low initial catches; Developing: rapidly rising catches; Mature: catches reaching and remaining around their historical maximum; Senescent: catches consistently falling below the historical maximum; Recovering: catches showing a new phase of increase after a period of senescence.)
SOURCE: FAO 2005.

fisheries. These events all combine to focus intense interest on the possible ecosystem-level effects of fishing and whether these influences can be addressed in fisheries management.

Fisheries by their very nature are extractive enterprises and therefore reduce the biomass of exploited populations. But sometimes the level of harvest has been too great, and scientists have shown that some stocks have been reduced to low fractions of their unfished levels. This is particularly evident in intensively fished coastal regions like the North Sea, the Gulf of Thailand, and the North Atlantic (Gulland 1988, Pauly and Maclean 2003, Garcia and Grainger 2005). But researchers are also now raising new concerns about the multiple impacts of fishing not only on target stocks, but also on other ecosystem components through bycatch (removal of non-target species), indirect effects (the removal of one species leading to the benefit or detriment of another), and habitat impacts (Dayton et al. 1995; Pauly and Christensen 1995; Pauly et al. 2002, 2005; National Research Council [NRC] 2002; Chuenpagdee et al. 2003; Pauly and Maclean 2003).

Furthermore, and possibly more important to this report, the public has been increasingly concerned that activities associated with fisheries have global impacts beyond those related solely to depletions of local stocks. This perception has been fueled by a series of highly publicized journal articles and reports (e.g., Pauly et al. 1998a, Jackson et al. 2001, Myers and Worm 2003). These papers suggest that fishing wild populations of marine fish and invertebrates has fundamentally altered the structure of marine ecosystems, resulting in severe depletion of populations at high trophic levels and propagating through whole communities of interacting species through indirect effects. However, these papers have not been without criticism. Subsequent articles have disputed the findings of these papers, and others have disputed the implications (or the broad application) of the conclusions presented, while still others continue to provide additional analyses. These ongoing discussions in the scientific literature are not as well known outside the fisheries science community as the papers mentioned above, yet they are equally important for deciding a course for future management. This report strives to present and discuss the related scientific literature by putting the range of perspectives in context and weighing the collective evidence on fisheries-induced ecosystem change.

## POLICY CONTEXT

Fisheries management has traditionally focused on the status of individual fish stocks; both the United States and the United Nations have policies regarding rebuilding overfished species. Recently, however, concerns have been raised about whether these approaches can account for the possible broader ecological impacts of fishing. As ocean management begins to embrace ecosystem-based principles, what are the specific concerns for fisheries? The possible ecosystem effects of fishing encompass a wide range of biological interactions including changes in predator-prey relationships and trophic dynamics (Hughes 1994, Pauly et al. 1998a, Steneck 1998), reductions in non-target species through bycatch (D'Agrosa et al. 2000, Lewison et al. 2004a), "cascading" effects mediated through food-web interactions (Frank et al. 2005, Ware and Thompson 2005), and the loss or degradation of habitats due to fishing pressure (NRC 2002). Fishing activities are also size and species selective, potentially resulting in changes to the genetic makeup of a stock as well as the structure of fished populations and communities.

Many of these effects are indirect and can alter complex interactions in marine communities, ultimately affecting the functioning of the ecosystem itself. However, fisheries impacts are occurring at the same time as other large-scale influences, such as El Niño events, other decadal-scale oscillations, and long-term climate change. In addition, a range of other human actions are also taking place in marine and coastal environments, resulting in the loss of wetlands and coral reefs, eutrophication, and pollution. Isolating the underlying cause of ecosystem effects is extremely difficult. All of these actions could be occurring individually or in concert to drive a relatively pristine marine ecosystem to one that is fully utilized[1] and, eventually, to one that is degraded.[2] Ultimately these effects, individually or collectively, may result in shifts in marine ecosystems that may or may not be reversible.

Another issue at hand is how to manage and protect marine ecosystems when goals and actions are based on incomplete understanding of ecosystem components and what constitutes an undisturbed ecosystem. Some archeological, paleo-ecological, and historical data analyses indicate that fishing by humans has altered the structure and function of marine ecosystems for centuries or millennia (e.g., Jackson et al. 2001). Such changes would have predated modern descriptions of ecosystem structure and will affect conclusions about why some ecosystems have undergone dramatic changes, perhaps shifting the scale on which ecosystem alteration is measured.

Many scientists and managers agree that a well-informed discussion is needed on what we consider the "ideal" state of the ocean to be, setting in motion thoughtful plans about how to achieve our common goals. Many questions need to be answered to reset the course of fisheries management: How much rebuilding is enough or desirable? Do levels of productivity observed for the last 50 years provide an adequate baseline for setting current goals? What tradeoffs are needed among species or among uses to allow management of the ecosystem as a whole? Some of these questions are scientific in nature; others are societal choices that need to be made in the public-policy arena.

---

[1] In this report, utilized is intended to describe functioning ecosystems that can sustain numerous uses, including fishing.

[2] Degraded is used to describe ecosystems that have been exploited to a point where there is a loss of desired uses, including a reduction in overall productivity or the loss of species.

## SCIENTIFIC CONTEXT

Ecosystems are inherently complex and any disturbance, such as removals of target species by a fishery, is likely to affect other components of the system; however, assessing these effects is often extremely difficult. Science continues to elucidate these interactions, but gaps in scientific knowledge ultimately limit the ability to fully assess the impacts of fisheries on marine ecosystems and to anticipate the likely response of interacting food webs to fishing.

Confounding the issue is the lack of independent and comprehensive fishery data in most regions, leaving fisheries landings and fishing effort data as the only indicators of ecosystem change in some cases. However, the conclusions that can be drawn from fishery-dependent data are often controversial because fisheries change in response to many factors (e.g., markets, management regulations) in addition to alterations in the ecosystem that they exploit. Alternative interpretations of fishery data sets are possible depending on how the analyses treat biases and limitations in the primary data and on the assumptions scientists make to fill the gaps in available information.

Further, consistent data going back more than 20 to 30 years are often difficult to obtain, a problem that makes characterizations of past conditions as tricky as predicting the future of fished ecosystems. Often, less formal sources of information are employed to reconstruct the history of fisheries and ecosystems prior to the implementation of regular monitoring programs (e.g., Rosenberg et al. 2005). Clearly, these studies are important for understanding the likely impacts of fishing operations on marine ecosystems, but this evidence alone is often inconclusive.

Scientists' ability to qualitatively and quantitatively model ecosystem dynamics and predict complex ecosystem interactions has rapidly advanced in recent decades. Continuous improvements to species-interaction models, energy-balance models, and ecosystem models have raised the possibility of applying such methods to management. In fact, some have evolved to the state where they have been used on small scales to enumerate policy options (Christensen 2005, Kitchell 2005). And while other models are still in the development stage, the possibility remains that models can be used to construct future scenarios of various ecosystem effects based on initial input conditions including biologic, economic, and management parameters.

## POLICY CHOICES AND THE ROLE OF SCIENCE

Broad stewardship options are available for addressing the general effects of fishing on marine populations, food webs, and communities. As a starting point, one can envision some early dynamic state of a wilderness ocean. In a number of cases, humans began exploiting the living resources of the sea well before written records documented this early state. What now exists is an ocean of living resources that has been modified by fishing, overfishing, and other factors not directly linked to fishing. Whether this state is described as utilized or degraded differs among regions, fisheries, and perspectives. For example, in the United States, the North Pacific region is often referenced as the example of a productive, utilized system in which fisheries management has generally succeeded at controlling harvest rates, while the New England region exemplifies the opposite type of system, one with significant impacts due to overfishing. However, no locations likely remain in the world's oceans that are entirely free of human influence. And while achieving the wilderness state may be neither possible nor desirable, understanding what that state may have been, and how much change has taken place, is important for determining and setting future management goals.

Recognizing that fishing practices have changed the ocean, the obvious question presents itself: Where do we, as a society, want to go from here? Several broad policy options are available. The traditional option would be to continue managing fish populations with maximum sustainable yield (MSY)[3] serving as the target catch for each individual fishery. However, single-species management has not been successfully implemented in many cases, with many populations overfished as a result. Continuing to overfish would further reduce the productivity of the stocks and make it more difficult to achieve a well-managed, utilized ecosystem.

On the other hand, the consequences of implementing single-species management more effectively than in the past are not easy to predict. It is possible that just eliminating overfishing would allow for some stocks or ecosystems to recover to former, more appropriately utilized states that meet societal goals. Another option would be to manage using more conservative fishing targets than those based on MSY. In fisheries managed by the North Pacific Fisheries Management Council, many species are harvested at levels defined by optimum yield,[4] well below MSY. However, this approach is often not implemented in other regions because of pressure to raise catch levels to accommodate various sectors of the fishing industry. Alternatively, policy options could incorporate multiple-species management strategies based on analyses of species interactions, food webs, and community dynamics. For systems that are already overutilized or potentially degraded, this approach might be termed ecosystem rehabilitation, *sensu* Francis et al. (1979). The basic premise of ecosystem rehabilitation in a fisheries context is that a reduction in fishing pressure, informed by knowledge of species interactions, would allow recovery to a less disturbed state. How closely the ecosystem could approach some previous ocean condition would depend to some extent on the magnitude of the fisheries management controls implemented. But, conceptually,

---

[3]See glossary (Appendix D) for definition.
[4]See glossary (Appendix D) for definition.

these actions may move the ecosystem closer to that condition, perhaps allowing society to take greater advantage of the overall productivity and to harvest the desired mix of living resources.

However, even for well-managed fisheries, a larger question remains. If managers were able to implement specific management actions to achieve a desired point on the scale from pristine to degraded ecosystems, how would this point be chosen? Commercial fishing, tourism, recreational fishing, and many other uses affect, and are affected by, the state of the marine ecosystem. Each stakeholder group will most likely desire a different mix of resource uses and different outcomes from management activities. In addition, the oceans provide large-scale ecosystem services, such as oxygen generation and nutrient cycling. How are these "values" accounted for when deciding ecosystem goals?

Without the ability to satisfy all constituents, the management of natural resources usually results in a series of tradeoffs between various user groups and between different ecosystem services. For example, one easily defined tradeoff would be associated with the recovery of a single stock. Catch rates would be lower in the short term; however, they presumably would be higher once the stock increased. More generally, tradeoffs involve multiple resources and ecosystem services, and they may create conflicts between different user groups with competing goals. The recovery of a top predator, for example, may negatively impact a fishery based on its preys.

When different users are in conflict (i.e., one use precludes or impacts another), the tradeoffs between uses require resolution of public policy decisions and value judgments. But both natural and social science have an important role in informing management decisions by revealing the range of potential outcomes based on a more complete understanding of ecosystem functioning, human behavior, and the connections between interacting species.

## MOVING TOWARD ECOSYSTEM-BASED MANAGEMENT

Both the U.S. Commission on Ocean Policy (2004) and the Pew Oceans Commission (2003) stressed the need to move away from sectoral management of the oceans (e.g., fisheries, shipping, water quality, oil and gas, invasive species, critical habitat, protected species, etc.) and toward an ecosystem-based management approach to ocean and coastal resources. The statement of task for this study was specifically to examine the ecosystem effects of fisheries and to consider the implications for fisheries management. The study's recommendations could be implemented by the National Oceanic and Atmospheric Administration (NOAA) National Marine Fisheries Service (NMFS) and the Regional Fisheries Management Councils to facilitate (1) a move from singles-species management to multi-species management, (2) consideration of marine food webs, and ultimately (3) ecosystem-based fisheries management. But it is essential to make the distinction between these fisheries-focused management decisions and true ecosystem-based management.

Many sectors other than fisheries impact marine ecosystems (Breitberg and Riedel 2005), and many different regional, federal, state, and county agencies have legislative responsibilities relating to marine systems. For example, decisions with respect to oil and gas extraction and shipping can have impacts on fisheries or protected species, but selection of sites for oil and gas development is not the responsibility of fisheries managers. Furthermore, water quality management—the responsibility of yet another agency—begins in the uppermost reaches of the watershed and has impacts downstream in the coastal zone, potentially affecting fish health or growth, and limiting viable habitat (Craig et al. 2001).

This report outlines the potential ecosystem effects of fishing and discusses the potential role of larger ecosystem impacts in fisheries management planning. But these actions will be only a part of a comprehensive ecosystem-based approach to management. As stated by the U.S. Commission on Ocean Policy (2004), the implementation of ecosystem-based management demands the active involvement of multiple agencies, requiring substantial, perhaps unprecedented, cooperation among management agencies at a number of levels of government and across issues.

## COMMITTEE APPROACH AND REPORT ORGANIZATION

The National Research Council (NRC) Committee on Ecosystem Effects of Fishing: Phase II—Assessments of the Extent of Ecosystem Change and the Implications for Policy was charged with reviewing and evaluating the current literature on the impacts of modern fisheries on the composition and productivity of marine ecosystems (see Box 1.1). NMFS, the study sponsor, asked the committee to discuss the relevance of these findings for U.S. fisheries management, identify areas for future research and analysis, and characterize the stewardship implications for living marine resources. The committee took the approach of reviewing the current literature to provide a larger context for the findings of several widely publicized studies and to evaluate whether the weight of the collective evidence is sufficient to justify changes in the U.S. approach to fisheries management. The findings and recommendations of the committee were based on presentations heard at three public meetings (see Appendix C for meeting agendas), published literature, and their own expertise. This report examines the current scientific evidence for ecosystem effects of fishing, including changes in abundance, biodiversity, and genetic structure of populations; food-web effects such as trophic cascades and species interactions; and both physical and fishery-induced regime shifts (Chapter 2). Subsequent chapters discuss how these kinds of effects might be addressed by changing how the United States manages fisheries (Chapter 3) and how interactions among species, uses, sectors, and values could be accounted for in fisheries management decisions (Chapter 4).

# INTRODUCTION

> **BOX 1.1**
> **Statement of Task**
>
> Recent high profile scientific reports suggest that there have been fundamental changes in marine ecosystems as a consequence of large-scale, global fishing activities. The authors have used historical data sets, meta-analytic techniques, and population models to "hindcast" the abundance of marine species before the advent of modern fishing activities. Several conclusions from these studies have received considerable media coverage and raised public concern and controversy over the effects of fishing on marine ecosystems. Examples of these conclusions include: (1) fishing has typically reduced the abundance of large predatory fish stocks by 90 percent; (2) fisheries have been "fishing down the food chain" by successively depleting stocks from top predators to grazers; and (3) focus on modern trends in abundance without regard to preexploitation conditions results in "shifting baselines" that set targets for recovery that are too low relative to the potential productivity of the ecosystem.
>
> This study will review and evaluate the current literature on the impacts of modern fisheries on the composition and productivity of marine ecosystems. The report will discuss the relevance of these findings for U.S. fisheries management, identify areas for future research and analysis, and characterize the stewardship implications for living marine resources.

Chapter 5 discusses the research needed to better understand multi-species interactions and the social and economic science needed to improve management strategies. Chapter 6 contains the recommendations of the committee, setting forth a research and stewardship agenda that will support a more holistic approach to fisheries management. Primary emphasis is placed on U.S. fisheries management—taking into account that much of the existing literature is global in nature and may not apply to conditions within the U.S. exclusive economic zone. Phase I of the NRC series on the ecosystem effects of fishing, *Effects of Trawling and Dredging on Seafloor Habitat* (NRC 2002), considered the habitat impacts of these fisheries on the seafloor and therefore will not be discussed in this report.

# 2

# Evidence for Ecosystem Effects of Fishing

Recent scientific literature has raised broad and multiple issues concerning changes to marine populations and food webs caused by fisheries removals. Vitousek et al. (1997) argue that no ecosystems on earth, including those in the ocean, are "free of pervasive human interference." Here the concern is with the influence of fishing. Yields from ocean fisheries are approaching their upper limits (Botsford et al. 1997), and with the total global harvest of marine capture fisheries[1] reaching 84.4 million metric tons in 2002, the likely maximum potential of conventional target species appears to have been reached (FAO 2005, Garcia and Grainger 2005). One can assume that removals of this magnitude must have some appreciable impact on ocean ecosystems; but to what extent has the species composition and biodiversity of the ocean changed as a result of fishing? And to what extent have these changes altered the current and potential benefits from the ocean as well as the functioning of ocean ecosystems?

Traditional fisheries management has been predicated on biomass reductions to 30 to 50 percent of unfished levels to maximize production (Mace 2004). Therefore, the fact that total biomass of fished species has decreased over time is not surprising. Yet, fishing is both size selective and species selective, meaning that the abundance and mean size of fished species are often reduced, and the genetic structures of populations are potentially altered. Furthermore, species interactions are often complex and fishing can modify elaborate connections in marine communities and food webs. These changes to populations and communi-

---

[1] Capture fisheries do not include production from aquaculture.

ties can and do alter species interactions and the functioning of ecosystems. The relations between "cause" and "effect" are often, perhaps always, non-linear and may include shifts in the state of the whole ecosystem.

In this chapter, the mechanisms, evidence, and magnitudes of fishing's effects on marine ecosystems through modification of populations and food-web structure and function are reviewed and evaluated. Topics addressed are: (1) changing the abundance of fished stocks and species groups, altering biodiversity, and changing the genetic structure of populations; (2) altering food-web structure and function through the dynamics of trophic cascades; (3) fishing down and through food webs; and (4) inducing regime shifts through either physical or biological forcing. This chapter also presents a discussion of the reversibility of fishery-induced changes and the possible time frames for recovery.

## CHANGES IN ABUNDANCE AND BIOMASS

Declining biomass is an expected effect of fishing on populations and is necessary for the density-dependent increase in production that is the basis for sustainable fisheries harvests, but in many cases overfishing has resulted in the collapse of populations and the fisheries that depended on them (e.g., northwest Atlantic cod). Numerous papers point to the decline in food fish biomass in various areas: the North Atlantic (Christensen et al. 2003), West Africa (Christensen et al. 2004), southeast Asia (Christensen 1998), the Gulf of California (Sala et al. 2004), and broadly around the world (Gulland 1988, Pauly and Maclean 2003, Garcia and Grainger 2005). The following point, drawn from Hilborn et al. (2003, p. 368), is replicated frequently in the fishery literature:

> United Nations (UN) Food and Agriculture Organization's (FAO) estimate that "75 percent of the world's fisheries are fully or overexploited" has been widely quoted. Considering that being fully exploited is the objective of most national fishery agencies (and therefore not necessarily alarming), of more concern is the estimate that 33 percent of the U.S. fish stocks are overfished or depleted.

For U.S. fisheries, the pattern of overfishing among regions and stocks is characterized by heterogeneity (National Marine Fisheries Service [NMFS] 2005). In the same waters, some stocks are overfished while other stocks are not. The same species can be overfished in some areas while not in others. The proportion of stocks that are overfished or are experiencing overfishing varies greatly among U.S. management areas. The average for known stocks in 2004 was about 28 percent overfished, but proportions among regions based on at least 25 stocks ranged from 10 percent to 44 percent of the stocks. The lesson to be learned from this is that effects of fishing on exploited stocks vary greatly among and within regions. A danger of overgeneralization is always present. The depletion pattern is spatially heterogeneous, both in U.S. fisheries and worldwide.

# EVIDENCE FOR ECOSYSTEM EFFECTS OF FISHING

Myers and Worm (2003) analyze the worldwide decline in longline catch per unit effort (CPUE) of predatory fish communities brought about by industrialized fisheries. Their analyses of 13 oceanic and coastal fisheries include the tropical, subtropical, and temperate Atlantic, Pacific, and Indian Oceans; the Gulf of Thailand; Saint Pierre Bank; the Antarctic Ocean off South Georgia; and the Southern Grand Banks. Typical reductions in catch per unit effort in the longline fishery were 80 percent of original catches within 15 years of the onset of industrialized fishing (Figure 2.1). Their results have been widely quoted: "Using a meta-analytic approach, we estimate that large predatory fish biomass today is only about 10 percent of pre-industrial levels" (Myers and Worm 2003, p. 280). Further, they report that most newly fished areas showed very high catch rates,

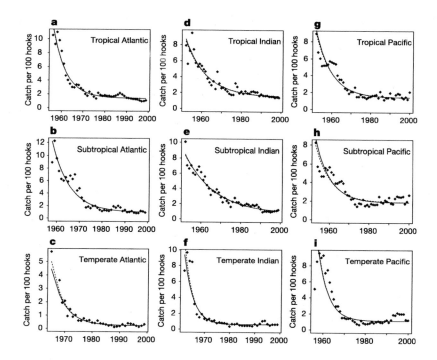

**FIGURE 2.1** Time trends show decreasing catch per unit effort for nine different oceanic ecosystems. In these open-ocean communities, catch rates fell from 6–12 individuals per 100 hooks down to 0.5–2 during the first 10 years of exploitation. Relative biomass estimates from the beginning of industrialized fishing (solid points) are shown with superimposed fitted curves from individual maximum-likelihood fits (solid lines) and empirical Bayes predictions from a mixed-model fit (dashed lines).
SOURCE: Myers and Worm 2003; reprinted by permission from Macmillan Publishers, Ltd.

but declined to low levels after a few years, resulting in abandonment of once-productive areas.

Others have published less extreme estimates of biomass declines in the north central Pacific based on production models. For example, Cox et al. (2002a) estimate that between 1950 and 1998, biomass declined to 21 percent for blue marlin and 56 percent for swordfish. Conversely, estimated biomasses of juvenile bigeye (*Thunnus obsesus*) and yellowfin (*T. albacares*) tuna increased to 112 and 129 percent, respectively. Based on all their observations, Cox et al. conclude that the changes generally represent decreases in top predators and subsequent increases in small tunas, on which the top predators prey.

The changing abundance of tunas and billfish as a consequence of fishing is a subject of active analysis and the last word is likely not in. Some question the analyses and the magnitude of decline estimated by Myers and Worm (Walters 2003, Hampton et al. 2005, Polacheck in press), and many tuna and billfish managers note stable or increased catches of some species even after the alleged collapses. For example, Walters (2003) argues that Myers and Worm overestimate declines because they ignore unfished time and spatial cells in the database; he postulates that because the Japanese longline fishery was initially concentrated in a small area, the initial drops in catch per unit effort were not reflective of overall stock trends, an example of a phenomenon known as hyperdepletion (Hilborn and Walters 1992). In his alternative analyses, declines still occur, but were similar in one analysis and not as large in another. Hampton et al. (2005) argue that changes in catch per unit effort used by Myers and Worm are not a reliable indicator of abundance because fishermen target certain species in response to market demands and price and, additionally, the equatorial Pacific is not included in their analyses. When this area is included, the declines in CPUE in the western Pacific are 70 percent over 50 years for yellowfin tuna, but stable for bigeye tuna since about 1960, even though total catch has continuously increased. In their rebuttal, Myers and Worm (2005a) counter these criticisms and argue that their estimates of declines in large predators are conservative.

Polacheck (in press) makes a number of significant points in analyses of Indian Ocean data that suggest that the Myers and Worm estimates of declines in abundance are too high and that the apparent collapses are inconsistent with the increases observed in the pelagic fisheries over all years. Based on new analysis, Polacheck agrees that a rapid decline in longline catches occurred during the 1950s and 1960s, resulting in catch rates that were 20 percent of the early values by the early 1980s (Figure 2.2). But, in the 1980s, total catches, summed across all types of gears (longline, purse seine, and others), began to increase and by 2000 were two to six times greater than in the early 1970s. During this period of increased catches, the CPUE did not decline; in some species it even increased slightly. The growth in yield of these fisheries is due mostly to the introduction of new gear types—primarily purse seining, which targets smaller individuals nearer the surface than those caught by longlines—and to the introduction of existing

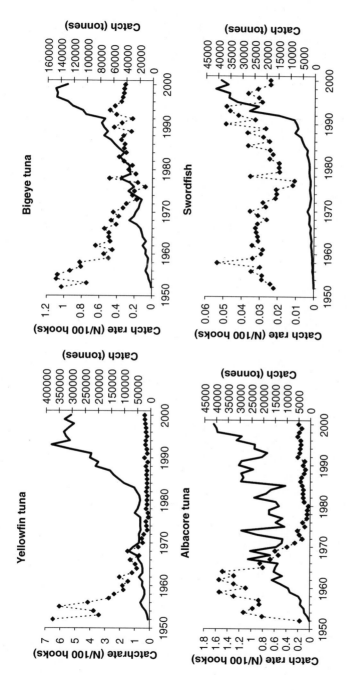

**FIGURE 2.2** The annual nominal catch rates by Japanese longliners (dotted line) and estimates of total catch from all fisheries (solid line) are shown for the four principle species of tuna and billfish caught by longliners in the Indian Ocean. The increase in total catch is inconsistent with the hypothesized stock collapse indicated by the decreasing catch per unit effort.
SOURCE: Reprinted from Polacheck in press, with permission from Elsevier.

fisheries to new geographical areas. The observed increasing catch and stable catch per unit effort are inconsistent with collapsed stocks, even though the larger fishes captured in the longline fishery had been reduced in abundance. Apparently, the early declines in longline catches did not drive total population declines as these populations were later able to support fisheries targeting smaller individuals with different gears. As Cox et al. (2002a) pointed out for the Pacific, the declines of the largest fish were more than compensated by the increased production and capture of the prey of these large individuals. Production likely increased because cannibalism by and competition with the larger individuals was reduced, and the fishery was then able to take a higher proportion of individuals from slightly lower trophic levels.

Polacheck (in press) also was concerned that the Myers and Worm analyses, where they summed across all species in the Indian Ocean, had some biases as well. Myers and Worm did not point out that the species composition of the catch had changed over the 50 years, largely as a result of targeting different species for different markets. Finally, the spatial pattern of decline analyzed by Polacheck does not support the idea of serial depletion following the first years of exploitation in an area. Catch rates in newly fished grid cells were not greater than those in previously fished cells.

One can deduce from the subsequent analyses that the contentious findings in the Myers and Worm paper center mostly on large pelagic species; few argue that their conclusions do not represent the current situation for demersal fish included in the study. Also, as with all large-scale syntheses, their conclusions fail to realize the diversity of possible abundance declines when specific species and specific locations are examined. It seems likely, when these subsequent papers are taken into account, that the 90 percent decline reported by Myers and Worm (2003) is an overestimation. Yet even these newer papers report declines in the 65 to 80 percent range. Although the estimated *magnitude* of the decline may vary, all of these analyses indicate that *abundances* of large predators have been reduced. Polacheck (in press) points out that catches are, in fact, high and that the ecosystem effects of removing so many large predators have not been evaluated nor are the data being collected to evaluate such effects. He also warns that as the debate continues on the meaning of the sharp declines in catch per unit effort in the early years of these fisheries, it would be unwise not to respond to clear evidence of overfishing of some of these tuna stocks in recent years.

## Biomass Recovery

As a response to reported declines, some analyses of worldwide data have been used to examine the ability of depleted stocks to recover to former abundance levels. After analyzing trends for 90 stocks, Hutchings (2000) finds little evidence for rapid recovery from prolonged declines, with the exception of early-maturing herring and related species. He concludes that although the effects of overfishing on single species may, in general, be reversible (Myers et al. 1995), the actual time required for recovery appears to be considerable and species-specific. However, as Mace (2004) points out, Hutchings's analysis does not consider whether or not there had been a reduction in fishing pressure after the stock declined. Indeed, stock recoveries were more likely to occur when exploitation was reduced (Hutchings and Reynolds 2004) or when stocks had not been severely depleted. But even in cases of reduction in fishing mortality, populations that have been driven to extremely low levels exhibited little or no recovery after 15 years. More recent data not included in this analysis may be more encouraging, as some stocks in New England show recent increases in biomass (Mace 2004). However, even if fishing mortality is reduced, recovery may be slow or may fail. The most well-known case is that of the northern Canadian cod, which has not rebounded after 15 years of low or no fishing and is unlikely to do so in the near future (Bundy 2005, Bundy and Fanning 2005).

## Biodiversity and the Extinctions of Species

Biodiversity is intricately related to both community productivity and stability (Tilman 1996, Worm and Duffy 2003). Reductions in biodiversity could mean that the ecosystem is more vulnerable to changes that previously could be absorbed. Changes in species composition, species richness, or functional type also affect the efficiency with which resources are processed within an ecosystem, meaning that the biogeochemical functioning of an ecosystem might also be impaired by a loss of species richness (Naeem et al. 1999). At the very least, evidence reveals that the likelihood of fishery-induced regime shifts (discussed later in the chapter) may increase when humans reduce ecosystem resilience by decreasing stock size and altering size structure through fishing activities.

Worm et al. (2005) report quantitative declines in diversity for heavily fished areas as measured by species richness (expected number of species) and species density (species per unit of fishing effort) (Figure 2.3). For longline catches in the Atlantic and Indian Oceans from 1950 to 1999, richness declined by about 10 percent. For the Pacific, there was an initial decline, but levels in the 1980s and 1990s returned to those previously seen in the 1950s. Declines in species density were greater, about 50 percent in the Atlantic and Indian Oceans and about 25 percent in the Pacific Ocean. Such declines in diversity are significant, but they are less dramatic than the proportional declines in abundance reported by Myers and Worm (2003).

High diversity areas were mostly in the subtropical seas. In the Worm et al. (2005) analysis, the two measures of diversity were related to sea surface temperature (greater at intermediate latitudes), sea surface temperature gradient (greater in high gradient, frontal areas), and dissolved oxygen (lower with low dissolved oxygen). An El Niño signal was also documented in the interannual changes in diversity. Furthermore, the slopes of temporal decline in the two

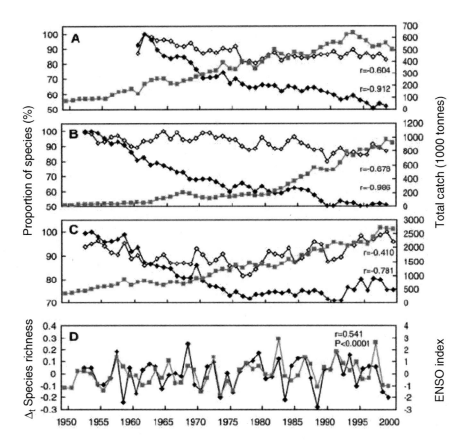

**FIGURE 2.3** Proportion of species (in percent of maximum) per 50 individuals (species richness, open diamonds), per 1000 hooks (species density, closed diamonds), and total catch (squares) of tuna and billfish across the (A) Atlantic, (B) Indian, and (C) Pacific Oceans. Species richness (D) shows pronounced year-to-year fluctuations and decadal declines of 10 to 20 percent in all oceans (black line shows species richness in the Pacific; gray line shows El Niño Southern Oscillation [ENSO] index), a trend that reverses in the Pacific in the 1970s. Species density shows gradual declines, ~50 percent in the Atlantic and Indian Oceans and ~25 percent in the Pacific. These declines were most pronounced in intensely fished tropical areas, particularly in the Indian and Atlantic Oceans.
SOURCE: Reprinted from Worm et al. 2005, with permission of the American Association for the Advancement of Science. © 2005 AAAS.

measures of diversity were statistically related to a five- to tenfold increase in fishing effort, and the intensity of fishing was concluded to be a significant cause.

Of the five threats to marine biodiversity, Jackson et al. (2001) place fishing effects first both in magnitude and in time. In most coastal ecosystems they evaluated, fishing impacts occurred first and then were followed in time by pollution, habitat destruction, invasive species, and finally climate change. But as they rightly note, the fact that many overfished species still exist, even if they are commercially extinct, offers the potential for restoration of ecosystem structure and function if fishing and other factors can be controlled. This potential provides opportunities forgone in many terrestrial systems where the megafauna have been forever lost or reduced to an irrecoverable remnant. But can reductions in abundance from fishing lead to irrecoverable extinctions? Further, what kinds of species are more likely to go extinct and are there specific areas more susceptible to extinction owing to fishing?

Jackson et al. (2001) note that few species of marine megafauna are globally extinct. Only 12 global extinctions of marine species have been documented, including 3 mammals, 5 birds, and 4 invertebrates (Carlton et al. 1999); however, many species are now reduced to such low population levels that they are ecologically or commercially extinct. For example, a number of species of elasmobranchs (14 in the Gulf of Lions in the Mediterranean and 9 in the Bay of Biscay) have disappeared completely in trawl surveys (Aldebert 1997, Quero 1998). Examining the relationship between species extinctions and fishing, Dulvy et al. (2003) published a compilation of 133 known local, regional, or global extinctions of marine populations and reported that exploitation was the cause in 55 percent of these cases. Examples from their list, representing either local or regional extinctions due to exploitation, include the sea otter (Carlton et al. 1999, Day 1989), European sturgeon (Jukic-Peladic et al. 2001), angel shark (Quero 1998), and bumphead parrot fish (a decline that they state was due to the effects of an artisanal fishery). But even the authors note that the quality of evidence supporting some of the entries on their list is variable.

Often the bycatch species of the target fishery are those most in danger of precipitous declines since fishing effort is not influenced by their abundance. Bycatch increases the extinction probabilities for a growing number of non-target species. For example, net entanglement poses a major threat to the seriously endangered northern right whale, accounting for a quarter or more of annual mortalities (Kraus et al. 2005). Pacific populations of leatherback turtles and many albatross are also in steep decline owing to unsustainable mortality from longline fisheries (Lewison and Crowder 2003, Lewison et al. 2004b).

Using a simple population model, Worm et al. (1999) and Myers and Worm (2005b) estimate the sensitivity of species to extinction and the proportions of populations going extinct under various rates of fishing mortality. Not surprisingly, in their model, sensitive species with lower reproductive rates (fewer young and later age of maturity)—such as sharks—are twice as likely to go extinct as bony fishes. But even bony fishes have high extinction probabilities at high rates of fishing. Modeled extinction rates are also greater when recruitment into the fishery occurs prior to maturity. These modeled results match general expectations and suggest that future extinctions may result from high rates of fishing. Hutchings and Reynolds (2004) argue that IUCN (The World Conservation Union) marine-fish extinction criteria (IUCN 2001) should not likely differ from terrestrial plants and animals and freshwater fishes. They back up this statement using analyses of empirical data on recovery probabilities for severely depleted populations. Dulvy et al. (2003) also conclude that high fecundity does not necessarily protect fish species from extinction risks, as has been previously believed.

While there have been few global extinctions as a result of fishing, the potential exists that future extinctions—including ecological extinctions—are possible, contrary to past thinking. The data presented for the declines in biodiversity and new examinations into the extinction potential for marine fish indicate that these issues deserve consideration in fisheries management decisions, especially as managers try to incorporate larger ecosystem concerns (the multispecies nature of fisheries management, accounting for effects of predation and competition).

## GENETIC CHANGES IN POPULATIONS

Law and Stokes (2005) suggest that fisheries are an enormous uncontrolled selection experiment—removing some classes of individuals in preference to others. This "experiment" is also continually revised in some fisheries as managers set new regulations such as size limits, catch quotas, and closed areas. These regulations influence the behavior of fishermen and lead to mortality of marine organisms of particular sizes, life histories, and behaviors. As long as there is an appropriate genetic component to variation in traits under directional selection, there can be no question that marine populations evolve as a result of exploitation.

New experimental evidence suggests that strongly size-selective fisheries may reduce the growth potential of individuals, leading to less productive populations. Conover and Munch (2002) show that fishing can drive phenotypic evolution of life-history traits using experiments with the Atlantic silverside (*Menidia menidia*). They applied selective removal of small or large individuals in replicate populations and found that within four generations large genetic differences had developed between the populations. One such difference was a faster rate of body growth when small individuals were removed; the total biomass yield from these populations was nearly twice that of populations from which large individuals were fished. Within a small number of generations, size-selective fishing could bring about observable genetic change in life history, placing evolutionary pressure on the potential production of a population.

Since the 1970s, many major exploited fish stocks have undergone large changes in their life histories. Cod, haddock (*Melanogrammus aeglefinus*), and pollack (*Pollachius virens*) stocks in the Northwest Atlantic, for example, have declined in age and length at maturity (Trippel et al. 1997). Since the early 1990s, these stocks have shown substantial declines in both age and size at maturity of approximately 20 percent. However, it is unclear whether these changes have a genetic basis. On the other hand, Olsen et al. (2004) show that fishing pressure did result in genetic changes for northern cod. Using probabilistic maturation reaction norms, they conclude that early-maturing genotypes are favored relative to late-maturing genotypes prior to the collapse of the stock in the early 1990s. They also suggest that this trend has begun to reverse due to decreased fishing pressure caused by the moratoria imposed in 1992.

Data has also been collected on Pacific salmon that indicate a decrease in mean size at age following years of selective fishing (Ricker 1981). For species such as salmon, where harvest often occurs during their spawning period, fishing pressure can select for such life history traits as time of spawning. The selection can be for either late or early spawners depending on the time of fishing. Fishing during the spawning period for other marine species can produce interesting results. In the case of the northeast Arctic cod, calculations suggest that the ancestral fishery on the spawning grounds was beneficial to the long-term production of the fishery since it selected for late maturation (Law and Grey 1989).

Studies like those cited above make it clear that fishing has the potential to alter life-history characteristics of populations; some of these changes are likely to be genetically based. Of course, fluctuations in the physical environment can have direct effects on life-history traits and the overall production of the fishery. These factors confound attempts to disentangle the contribution of genetic changes. However, new methods designed to distinguish temporal change in reaction norms from proximal effects of the environment support the theory that genetic changes are also taking place; results on size and age at maturation of northeast Arctic cod suggest a strong signal of change still remains in place (Heino et al. 2002). In general, it is proving increasingly difficult to account for large changes in the life histories of exploited fish on the basis of environmental factors alone.

## THE PHENOMENA OF SHIFTING BASELINES

In marine conservation, there is often a lack of reliable baselines—measures of the pristine state or knowledge of what existed in the past—upon which to base future effort levels and fishery restrictions (Pauly 1995, Dayton et al. 1998, Myers and Worm 2003). Recent publications based on paleoecological, archeological, and historical data suggest populations of marine organisms were once much larger than currently observed in kelp forests, coral reefs, and estuaries (Jackson et al. 2001). Many methods for assessing fish stocks use historical records to estimate average stock sizes prior to exploitation and current levels of stock depletion relative to those levels. However, a real problem with long data sets is often that both fishing technology and assessment methodologies change with time, breaking the time series into stanzas that tend to become "incompatible." This leads to analyses being undertaken on only part of the data. In many instances, stock assessments only begin when catch-at-age data are available (typically in the 1970s or 1980s) because the particular stock assessment methods used require these types of data. As a result, analytical historical reconstructions often do not cover the entire period of exploitation, even if catch and survey data go back several more decades (NRC 1998). The problem can be severe in fisheries that already were intensively exploited before the systematic collection of fisheries statistics began, as has been shown for North Atlantic cod (Rosenberg et al. 2005).

An analysis of the pre-European densities of green sea turtles (*Chelonia mydas*) and hawksbill sea turtles (*Eretmochelys imbricata*) in the Caribbean relative to current population sizes concludes that these endangered turtles once grazed seagrasses and sponges in numbers ranging from tens to hundreds of millions (Bjorndal and Jackson 2003). These numbers are difficult for contemporary sea turtle biologists and marine ecologists to fathom. Another possible case of a shifting baseline is that of the northeast Pacific kelp forests (Dayton et al. 1998). The distribution and abundance of the major structure-forming species, the giant kelp (*Macrocystis pyrifera*), are highly influenced by changes in ocean productivity at a variety of scales from storm events to El Niño/Southern Oscillation fluctuations to other decades-long regimes of the Pacific Decadal Oscillation. Thus, the kelp "baseline" is a dynamic one. Sufficient historical data to quantify the roles for the major consumers in these communities (e.g., sea otters, large fishes, urchins, abalones) before exploitation are absent. But human exploitation began in this system some 10,000 years ago and likely changed the abundance of these consumers. Jackson (1997) makes a similar case for coral reefs; most modern reef ecology focuses on the last 50 years, but reef systems in the Caribbean had undergone centuries of change and loss to the megafauna before scientists began careful observations or experiments.

In the absence of reliable data about historical trends in abundance, it is not uncommon that diverging interpretations of the available data or different methodologies lead to widely different estimates of historical population levels. Estimates from catch records[2] can vary from estimates based on fishermen's logbook data, survey information, or other systemically collected data that can be more qualitative. One such example is the historical reconstructions of the great whales. Fishermen targeted great whales around the world for centuries; some are now extinct, many are endangered or threatened. It is known from catch records of depleted whale populations that a large proportion of their biomass was removed by targeted fishing throughout the world's oceans (see Hilborn et al. 2003). Using

---

[2]Catch record refers to either the number or tons of animals harvested.

genetic analysis of neutral genetic variation, Roman and Palumbi (2003) estimate that the number of whales in the North Atlantic prior to whaling was an order of magnitude higher than historical estimates based on whalers' logbook data. Although they explored various assumptions that might cause the genetic analysis to overestimate the baseline numbers of whales in the North Atlantic, their most conservative estimates are still three to five times higher than historical estimates based on catch records. However, both the demographic and the genetic models used to reconstruct historical abundances depend on uncertain model inputs, and a unified analytical framework may be needed to try to reconcile both approaches (Baker and Clapham 2004).

Only a decade ago, when Pauly (1995) introduced the concept of shifting baselines, the absence of early data was only part of the phenomenon. He argued that each generation of scientists and managers accepts as a baseline the conditions in marine systems and particular fisheries stocks that they observe early in their careers, and they then document changes over their careers. The next generation does the same, resulting in a gradual shift of the baseline in terms of species observed and abundances. A sense of history is often absent over periods longer than a career. This hypothesis was recently evaluated by surveying over 100 people representing three generations of fishermen from Mexico's Gulf of California (Saenz-Arroyo et al. 2005). Fish populations have declined steeply in this region over the past 60 years and researchers probed the fishermen's perceptions of these changes. Compared to young fishermen, older fishermen named five times as many species and four times as many fishing locations as having once been abundant but now depleted. The older fishermen recalled catch rates on their best fishing days as much as 25 times higher than those reported by the younger fishermen. These changes have occurred within living memory, but young fishermen's perceptions of the state of the system differed dramatically from those of older fishermen.

A long-term view is not employed by scientists and managers either because early information is lacking or because they operate only on recent information. This precludes a vision to rehabilitate ocean ecosystems to earlier and perhaps more ecologically and economically valuable states. Taking a long-term view is an approach that can improve our perception of the world and how it works or might work (Magnuson 1990).

## ALTERED FOOD WEBS

Fisheries not only affect populations but also alter the energy flows and species interactions in marine food webs and communities simply because all fished species are components of food webs and interact with other species through predation and competition. Thus, any alteration of a stock biomass or size and age structure also alters food-web structure, energy flow, and species interactions as well as the strength of these interactions in marine ecosystems.

Some responses can be compensatory in nature. The following section defines and explains what food webs are and how they are viewed and analyzed, reviews information on trophic cascades in food webs, and considers top-down (consumer-control) and bottom-up (resource-control) effects on species.

## Food Webs and Trophic Interactions

Food webs are the road maps to known or even imagined species interactions and therefore display ecological connections. Their use as a descriptive tool extends back to the 1800s. Elton (1927), who termed them "food-cycles" and initially identified their central place in community ecology, used as his first example the North Sea web featuring diatoms, zooplankton, and copepods assembled by Hardy (1924). An unambiguous vocabulary has since evolved to describe their general features. "Links" connect individual predators to their prey; these are direct interactions. "Trophic level" has developed as a very general collective term, describing groups of species that are a similar distance in terms of energy transfers from the photosynthetic base.

A very simplified version of a food web is the ubiquitous food chain where smaller organisms are successively consumed by larger ones. Thus one reads of primary producers (organisms that are able to create biological energy through photosynthesis), herbivores, and primary and secondary carnivores.

Species feeding at more than one trophic level as adults (i.e., omnivores) are not easily categorized; assignment of a species to fractional trophic levels is increasingly common. Also, all fish feed at different trophic levels at different life stages. Due to their smaller size, juveniles are limited to feeding on small organisms, creating a situation where juveniles may be competing against species that will become their prey in later life stages. In more dynamical presentations of webs, a system of arrows scored by signs (+,-) can connect the entries. In this fashion it becomes possible to illustrate how a species can be "indirectly benefited," either through a reduction in predation or through reduced competition for resources. Analyses suggest that 40 to 50 percent of observed ecological impacts are transmitted through these indirect linkages (Schoener 1983, Wootton 1994, Menge 1995).

The notion of "strength of interaction" also can be added as a measure of how changes in prey and predator abundance affect the mortality rate of the prey, and how such changes affect mortality, growth, and recruitment of the predator. These can be "strong" or "weak" individual interactions (MacArthur 1972, Paine 1980) and can be presented on a per capita basis. Beyond characterizing species interactions as weak or strong, there are concerns about the functional and numerical responses of predators to changes in prey abundance. Are predators able to maintain their food intake rate at low prey densities, thus increasing predation mortality on depleted preys in a depensatory way? Or is prey vulnerability limited by refuges or by the predator switching to alternative prey resources?

Characterizations of food webs fall into three general categories, each with its own strengths and weaknesses (Paine 1980). Descriptive or classical webs can display the enormous trophic complexity of natural systems, but they treat all linkages as being equivalent (Figure 2.4). This visually appealing format can be replaced by a predator–prey matrix that quantifies specific dietary details including cannibalism. However, this format remains unable to incorporate indirect effects, although the potential pathways are illustrated.

A second food-web category emphasizes the flow of energy along connected links: Species can be identified as important or not by the magnitude of their linkages as conduits of mass or energy transfer (Figure 2.5). A limitation of this approach is that the magnitude of energy flow is not always a useful predictor of population responses and, therefore, the importance of a species may be underestimated. For example, quantitatively "weak" trophic flows from juvenile fishes (typically of small biomass) to their predators can represent large mortality rates for the juveniles with potentially large impacts on recruitment rates (Walters and Martell 2004). This is relevant to the choice of species to represent in food-web models, so that potentially important interactions are not ignored. Ecosim/Ecopath models (Polovina 1984, Christensen and Pauly 1992, Walters et al. 1997) widely employed in fisheries science fall into this category. These models are dynamical and assume that all energy entering the system must be accounted for (i.e., input is equated to output).

The final category, now called interaction webs (Menge 1995), allows critically significant or keystone species (Paine 1969), trophic cascades (Paine 1980, Carpenter et al. 1985), and the dynamically related swarms of indirect effects to be highlighted (Figure 2.6). Interaction webs are best used to distill an assemblage down to its known, most important ecological features and to identify indirect effects that develop as a consequence of apex predator manipulation. Their robustness depends on both substantial biological intuition and some form of system perturbation sufficiently intense to reveal before and after conditions. Therefore, identifying a baseline (and its natural fluctuations) can be of fundamental importance.

Although the entirely descriptive portrayals were historically important, and have been the focus of numerous attempts to distill general patterns from the static details, the two other approaches (energy transfer and interaction webs) are more realistic because they acknowledge that species interact with direct consequences for both participants. Tradeoffs in their respective utilities involve the consideration of space as a resource, the role of indirect effects, and the consequences of requiring that the models have a mass balance. In the energy/mass balance webs and models, both predatory and potential competitive interactions are effectively displayed. These are popular in fisheries ecology (e.g., Trites et al. 1999, Schindler et al. 2002) but have not yet been incorporated into scientific advice for fishery management.

Ecosystems are biologically complex, and food webs seem the most appropriate vehicle to display the implied interactions and their potential consequences

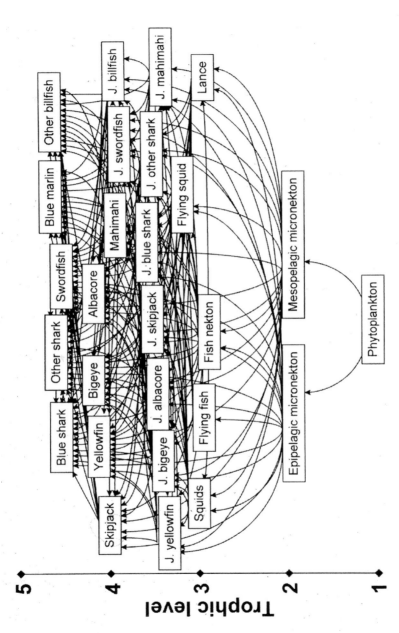

**FIGURE 2.4** An example of a descriptive food web for the central Pacific Ocean. Direct interactions are shown, but the strength of these interactions is not.
SOURCE: Kitchell et al. 1999.

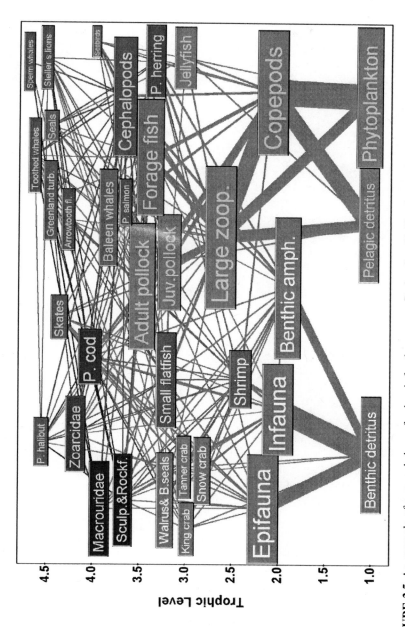

**FIGURE 2.5** An example of a mass balance food web for the western Bering Sea. Unlike the previous example, relative biomass of each species is indicated by the size of the box, and the strength of each interaction is represented by the width of the connecting line.
SOURCE: Reprinted from Aydin and Livingston 2003, courtesy of the National Oceanic and Atmospheric Administration.

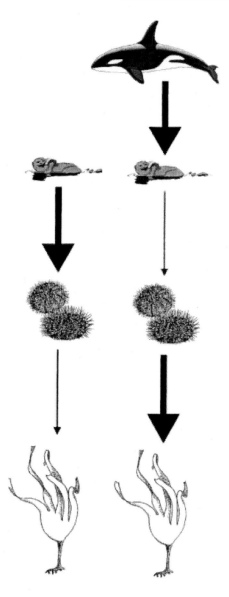

**FIGURE 2.6** Interaction webs show the strength of species interactions, where the per capita interaction strength is indicated by the width of the arrow, and how these interactions change due to perturbation. For this example, the perturbation is the introduction of orcas into a food web previously dominated by otters.
SOURCE: Modified from Estes et al. 1998, with permission of the American Association for the Advancement of Science. © 1998 AAAS.

in a fisheries context. Regardless of how they are visualized, food webs involve a necessary simplification by aggregating species, particularly those at lower trophic levels, into single trophic categories. No web displays all the known feeding relationships; usually, only those considered meaningful or common are included. Further, a problem with trophic models, particularly in data-poor areas, is that interactions are often developed using parameters estimated for different times and different places (e.g., from databases), and sometimes for different species, assuming that the trophic flow is constant in space and time—a very unlikely assumption.

Despite these and other constraints, webs based on energy flux have been broadly applied to exploited ecosystems (Walters et al. 1999; Pauly et al. 2000; Christensen et al. 2002; Cox et al. 2002a, 2002b; Martell et al. 2002; Olson and Watters 2003; Hinke et al. 2004; Shannon et al. 2004). Interaction webs, by featuring strong interactors and indirect effects, generate insights on predator-controlled ecosystems, especially in those experimentally tractable systems where space utilization must be considered. Can the approaches of interaction webs and energy flow webs be melded into useful fishery management protocols? This goal is highly desirable, especially when there are hints of significant benthic-pelagic coupling, such as on Georges Bank and in the Gulf of Maine (Witman and Sebens 1992), on the eastern Bering shelf (NRC 1996a), or across the entire North Atlantic (Worm and Myers 2003).

## TROPHIC CASCADES

Although humans have been fishing for millennia, the rapid expansion of industrialized fishing over the past 50 years has increased the impacts on marine food webs through both the magnitude of the removals and the exploitation of species in waters far beyond traditional fishing grounds. Removals of both target and non-target species (bycatch) can alter the abundance and productivity of components both higher and lower in the food web. But to what extent is the fishing effect limited to the population of the exploited species, and what additional effects cascade to the structure and function of the entire marine food web? It has been long accepted that the transfer of energy and thus organic matter upward through food webs influences community structure and overall productivity (Steele 1974). More recently, the importance of trophic cascades or the selective influence of consumers on the abundance and productivity of food-web components at lower trophic levels has emerged and is now recognized as possibly of equal importance in determining the dynamics of marine food webs (Estes et al. 1998).

Power's (1992) paper states the obvious: Both top-down and bottom-up processes must characterize all ecosystems. The interesting questions revolve around their relative influence. The notion that the removal of top predators could have dramatic effects on food-web structure and function actually began with the

Hrbáček et al. study (1961) of riparian fish ponds in Prague and with Paine's (1966, 1969) classic species-removal experiments in the rocky intertidal of the northeast Pacific. Data increasingly point to dramatic impacts on marine ecosystems caused by removal of large predators (Jackson et al. 2001).

Analyses of the statistical relations between different trophic levels in marine ecosystems provide evidence for both strong consumer effects (top-down) and strong influence of productivity at lower levels in the food web (bottom-up effects). Bottom-up productivity effects are apparent in comparisons among locations in the northeast Pacific (Ware and Thomson 2005). Strong cascading effects of removing top predators have been shown in the Scotian Shelf ecosystem off Nova Scotia using time-series data (Frank et al. 2005). (Both of these examples are described in Box 2.1.) In a comparison using freshwater lakes, Carpenter et al. (1991) estimate that the influence of consumer and nutrient effects on chlorophyll abundance are similar in magnitude; the error in the estimates of chlorophyll is lower when both top-down and bottom-up variables are included in the analysis. The joint influence of both consumer-driven and productivity-driven dynamics are not contradictory even if both are acting in the same ecosystem. The importance of each set of influences would be expected to differ among systems with respect to the magnitude of variability in production and the magnitude of variability in consumers over the time or space scales analyzed. Both could be equally important, or one or the other may dominate the dynamics.

Many other examples of predator-controlled cascades exist and they are more important to the discussion here than bottom-up influences since humans can be viewed as the top predator in many systems. These studies show that top-down control can affect every part of the system, with several of these examples ultimately affecting the foundational species, or habitat, of the ecosystem. In the Caribbean coral reef/sea grass system examined by Jackson et al. (2001), increased fishing leads to reductions in sharks and crocodiles and the extinction of Caribbean monk seals (*Monachus tropicalis*). Predatory fishes and invertebrates also decline due to fishing, as do grazers including manatees (*Trichecus manatus*) and sea turtles. The cascading effects propagate to the sea grass, sponges, and macroalgae (which increase) and to the corals (which decline). The current ecological role of grazing urchins, fish, sea turtles, and manatees is severely diminished and the primary producers are flourishing.

Silliman and Bertness (2002) use an interaction web to illustrate how a chain of coupled interactions potentially contributes to salt marsh grass (*Spartina*) destruction in the southeastern United States. Blue crabs (*Callinectes sapidus*) indirectly benefit the marsh grass by controlling the density of a marsh snail. However, these crabs are commercially valuable and their populations are declining. When snail populations were experimentally augmented, they reduced healthy *Spartina* stands to mud flats within eight months.

Perhaps the best known example of the unintended consequences of a trophic cascade linked to marine fishing/hunting—one that is relatively uncontaminated

> **BOX 2.1**
> **Bottom-up and Top-down Forcing in Marine Ecosystems**
>
> *Northeast Pacific—productivity control*
> Differences in productivity, measured using surface chlorophyll, in various locations in the northeast Pacific are reflected in the abundance of zooplankton and resident-fish yields (Ware and Thomson 2005). For areas off of British Columbia, the correlation between chlorophyll and zooplankton was +0.92 and the correlation between zooplankton and resident-fish yields was +0.87. This study provides strong quantitative, empirical evidence for bottom-up trophic linkages between phytoplankton, zooplankton, and resident-fish yields among areas at different spatial scales. Evidence of climatic decadal changes in the North Pacific influencing productivity throughout the ecosystem also comes from analyses of time series (Francis et al. 1998, Hare and Mantua 2000).
>
> *Scotian Shelf—predator control*
> The trophic cascade on the Scotian shelf was initiated by the virtual elimination of the ecosystem-structuring role of the large predators (Frank et al. 2005). In a trophic cascade where the consumers are driving the dynamics, adjacent trophic levels should be negatively correlated and those separated by an intermediate trophic level should be positively correlated. Alternatively stated, the prey of the large predators (i.e., small pelagic fishes and bottom-living invertebrates) should increase and show a negative correlation, which they did ($r = -0.61$ to $-0.76$), and the prey of the prey (i.e., herbivorous zooplankton) should then decrease. As expected, the correlation between the large predators and the herbivorous zooplankton, which are two levels apart, was positive, $r = 0.45$, and that between the herbivorous zooplankton and the phytoplankton (the next level down) was negative, $r = -0.72$. The final step in this cascade resulted in a decrease of nitrogen concentration that resulted from an increase in abundance of phytoplankton. This example provides strong quantitative, empirical evidence for a trophic cascade initiated by fishing down the large predators. The ecosystem response was nonlinear and resulted from complex interactions in a food web that included humans at the top.

by alternative explanations such as climate change, pollution, and habitat loss—is the ecosystem along the Aleutian Islands chain and North Pacific coast. The cascading relationship between sea otters (*Enhydra lutris*), kelps, sea urchins, other marine mammals, and fishing and hunting is a textbook example in marine ecology (Box 2.2).

Certain species exert strong controlling influences on marine food webs and these special species have an unexpectedly large impact on food-web structure and function. In other words, they have a large per capita effect. Such species are often found in top predator roles, which explains why systems are so highly altered when top predators are lost or severely reduced (Jackson et al. 2001).

## BOX 2.2
### Interacting Trophic Cascades in the Northern Pacific

Within the North Pacific food web, there is a direct correlation between sea otter and urchin populations—when sea otter populations decline, sea urchin populations increase owing to a lack of predation. When this happens, increasingly abundant sea urchins are able to virtually eliminate kelp populations from particular habitats. The alternate states of this community, kelps or urchin barrens, can persist for long periods of time as determined from the exploration of Aleut middens in Alaska (Simenstad et al. 1978). Because kelps are structure-forming species, they create habitat for many fishes and invertebrates; when kelps are lost, their associates are lost as well.

The balance between a sea otter- or sea urchin-dominated system changes spatially as well as temporally. After being protected from overhunting, the recovering populations of sea otters changed nearshore ecosystems by reducing the abundance of urchins and thus promoting kelp forest expansion (Estes and Duggins 1995). However, in the late 1990s, sea otter populations started to decline precipitously over large regions of western Alaska. The best explanation for these declines seems to be increased predation by killer whales (*Orca orcinus*) (Estes et al. 1998), the influence of still another food-web component. In an orca-dominated system, sea otters are suppressed, urchins recover, and kelp forests decline.

Populations of large pinnipeds, including Steller sea lions (*Eumetopias jubatus*), northern fur seals (*Callorhinus ursinus*), and harbor seals (*Phoca vitulina*) also collapsed in the western North Pacific beginning in the late 1970s. One interesting hypothesis suggests that the decline of marine mammals, including otters, in the North Pacific may be consistent with increased mortality, possibly from orcas, rather than from reduced food or any other bottom-up effect (Springer et al. 2003).

Springer et al. (2003) attribute the sequential declines in this suite of marine mammals (Steller sea lions, northern fur seals, harbor seals, and sea otters) to whaling in the North Pacific ecosystem. Killer whales likely consumed great whales (in fact, they were first dubbed "whale killers" by the early whalers [Scammon 1874]), and when the great whales were suppressed due to hunting, killer whales expanded their diet to include harbor seals, fur seals, sea lions, and, finally, sea otters.

---

Structure-forming species like kelps, sea grasses, and reef-forming corals are also strong interactors because they create a habitat that supports a high diversity of associated species.

This perspective of interaction in food webs is important to management. Humans have now become perhaps the strongest interactor in marine food webs. The previous examples and analyses point out that human policy decisions related to increasing or decreasing fishing (or hunting) pressures do occur and reverberate through trophic cascades in complex food webs and ecosystems. Sequences of events are set in motion and reveal themselves over decades. These changes have sometimes resulted in higher allowable catches of certain species that are released from predation, but in other cases the results are less favorable. Many of these cascades have been surprises and certainly were unintended consequences of management actions. However, methods are being tested to see if future cascades can be predicted. In three out of four simulations for the central Pacific Ocean, Cox et al. (2002b) were able to predict changes in small tunas and other major prey as the abundance of the larger Thunnids was reduced.

## FISHING DOWN AND THROUGH THE FOOD WEB

Long-term reductions in the mean trophic levels of fisheries landings have been measured worldwide. Explanations as to why trophic level changes in fisheries landings have occurred include (1) serial depletion of high-trophic-level species and subsequent replacement by lower-trophic-level species (i.e., the commonly held interpretation of fishing down the food web) (Pauly et al. 1998a), (2) sequential additions of lower-trophic-level fisheries while maintaining catches of higher predators (Essington et al. 2006), and (3) environmentally induced changes and natural cycles.

Pauly et al. (1998a) use FAO global fisheries landings statistics and estimates of trophic levels for fish and invertebrates determined by diet data and mass-balance trophic models. They conclude that the mean trophic level of fisheries landings worldwide have declined since 1950 (Figure 2.7). Globally, the overall mean decline in trophic level of fisheries landings was 0.05 to 0.10 per decade without the landings themselves increasing substantially. Further, most ocean areas were characterized as having declines in mean trophic level. The interpretation of the general nature of the steady declines was that they reflected a transition of the fisheries from long-lived, high-trophic-level piscivorous bottom fish toward short-lived, low-trophic-level invertebrates and planktivorous pelagic fish. Pauly's group also points out that as fisheries moved from high-trophic-level fish to lower-trophic-level fish, the catches initially increased as expected, but then subsequently declined. These patterns of decline were complex and resulted from several different causes. In addition, the group reported that a few FAO ocean areas either showed no clear trend or perhaps even an increase in trophic level (e.g., the IndoPacific); these were interpreted as inadequacies in the statistics.

Fishing down the food web has been argued to cause a reduction in the number and length of the pathways in the food web, making the ecosystems less resilient to environmental fluctuations (Pauly and Maclean 2003). Other possible implications include reduction of apex predator guilds, restructuring of marine ecosystems, and a loss of biodiversity. Yet it is unclear whether the changes impact overall productivity or simply divert the system productivity to other species. Fishing down the food web is often considered indicative of unsustainable fishing and has been viewed as the summary index of negative effects of fishing on marine ecosystems (Barange et al. 2004). According to Steneck (1998,

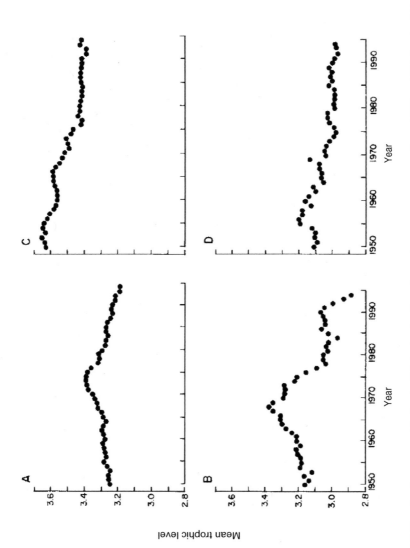

**FIGURE 2.7** Trends of mean trophic level of fisheries landings in northern temperate areas from 1950 to 1994 for the (A) North Pacific, (B) Northwest and Western Central Atlantic, (C) Northeast Atlantic, and (D) Mediterranean.
SOURCE: Reprinted from Pauly et al. 1998a, with permission of the American Association for the Advancement of Science. © 1998 AAAS.

p. 430), "Fishing down trophic cascades is the antithesis of sustainable harvests and is clear evidence of ineffective management." However, the cause of the reductions in mean trophic levels of fisheries is an active area of inquiry by the fishery and ecological science communities.

Caddy et al. (1998) criticize the analyses in Pauly et al. (1998a) based on four considerations: inadequate taxonomic resolution, the use of landings data as ecosystem indicators, aquaculture development, and eutrophication of coastal areas. These scientists do not disagree that a general decline in mean trophic level of marine landings was likely in many regions, but they are not convinced that the explanation is only "fishing down the food web." The concerns expressed by Caddy et al. (1998) are addressed by Pauly et al. (1998b) and later by Pauly and Palomares (2005). These responses include clarification of several misunderstandings, and in several cases they reanalyze the data or remove certain types of questionable data from the analyses. These changes do not alter their original conclusions, but the debate is likely to continue.

Given the potential utility of monitoring mean trophic levels of fishery landings in the eastern tropical Pacific, the Inter-American Tropical Tuna Commission (IATTC) estimates trophic levels for time series of annual catches and discards from 1993 to the present for purse seine fisheries that target tunas at high trophic levels (IATTC 2004) but also catch small quantities of sharks, billfishes, dorado (*Coryphaena* spp.), wahoo (*Acanthocybium solandri*), and other species. High mean trophic levels occurred when the annual bycatch of large apex predators (e.g., billfishes and sharks) increased. On the other hand, anomalies that resulted in lower trophic levels in the time series were caused by increased bycatch of manta rays and other species that feed on plankton and other small, low-trophic-level species. Variation in mean trophic levels through the time series thus reflects the relative composition of the bycatch, either higher catches of large apex predators or higher catches of lower-trophic-level species. This analysis again portrays the complexity of the causes in any change in the mean trophic level of a fishery over time.

A useful extension of the continuing analyses of the mean trophic levels of fisheries is developed by Essington and colleagues (2006), who determine the frequency of two alternative causes to explain the decline in mean trophic level of catches. The two patterns, already recognized by Sprague and Arnold (1972) and Pauly et al. (1998a), were distinguished as (1) fishing down the food web and (2) fishing through the food web (Figure 2.8). The first pattern, fishing down the food web, involves serial depletion of fished species from the top of the food web. The second, fishing through the food web, involves serial additions of lower-trophic-level species to the catches. Declines in mean trophic level by fishing through the food web can be caused by initial depletion of predators in the early stages of a fishery, followed by release of prey from predation, thereby increasing their abundance and their contribution to the fishery. An example of fishing through the food web provided by Essington et al. (2006) is from the

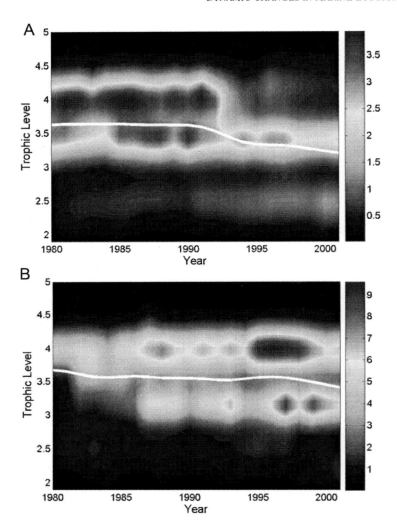

**FIGURE 2.8** Total yearly catch is shown for each 0.1 trophic level increment, indicated by the color bar on the right ($10^4$ kg yr$^{-1}$). The white line shows the mean trophic level. The Scotian Shelf (A) provides a typical example of the sequential collapse and replacement mode. The mean trophic level in fisheries landings exhibited a sharp decline from 1990 to 2001, owing to the collapse of the cod fishery, followed by a decline in the herring fishery and then the growth of the northern prawn fishery. In contrast, the Patagonian Shelf (B) exhibited a similar decline in mean trophic level over the same time period, but landings of high-trophic level species generally increased over this period, while new fisheries developed for lower trophic species.
SOURCE: Essington et al. 2006. © 2006 National Academy of Sciences, U.S.A.

Patagonian Shelf where the mean trophic level of the catch declined during the 1990s, but landings of a high-trophic-level species (Argentine hake, *Merluccius hubbsi*) increased during this period. The decline in trophic level of the catches from this region resulted from the addition of a new fishery for the shortfin squid (*Ilex argentinus*).

Essington et al. (2006) use annual catches by species and estimate the trophic level for each species to analyze trophic level of catches in 48 large marine ecosystems around the world as defined by Sherman (1991). Of the 48 ecosystems, 30 showed evidence of declining mean trophic levels, with an average decline greater than that described by Pauly et al. (1998a), who use more spatially aggregated data. Examples of both patterns were found, but sequential addition was more common than sequential depletion; in 23 of the 30 cases (77 percent), fishing through the food web prevailed. The sequential collapse and replacement mode was generally restricted to ecosystems in the North Atlantic, where management measures have been ineffectively implemented and many stocks have become grossly overfished.

One limitation that should be noted of both the Pauly et al. and the Essington et al. approaches derives from the use of catch data as an indication of stock status. In particular, the maintenance of high catches while a stock is being fished down may obscure a pattern of stock depletion. The example from the Patagonian Shelf is a case in point; while catches of the Argentine hake increased in the 1990s, the stock biomass decreased, reaching a critical status in 2000. These issues arise because most available data are fishery dependent, yet catch statistics are dependent on many factors including changing gear, effort, or area fished. This creates uncertainty about the extent to which the composition and trends in the catch reflect ecosystem changes.

Because both occur, the management implications of fishing down the food web (serial depletion of high-trophic-level species) and fishing through the food web (serial additions of low-trophic-level species) need to be addressed on a case-by-case basis. In particular, because fishing down and fishing through the food web affect multiple trophic levels simultaneously, nonlinear responses in the ecosystem may be more likely and managers may have to be sensitive to these possible changes. However, the overall message derived from examining both analyses is that it is not possible to determine the underlying force of change using only a trophic index. Therefore, while this index can be used to indicate that food-web level changes may be occurring, more data will be needed to judge the significance of the change to sustainable fishing.

## Trophic Efficiency of Food Webs

Irrespective of what causes the change in the mean trophic levels of landings, a reduction in mean trophic level of an ecosystem would be expected to result in an increase in productivity available for fisheries harvest. Transfer efficiency from one level of the food chain to the next averaged 10.13 percent (standard deviation 5.81) from analysis of 140 estimates of transfer efficiencies from 48 trophic models of aquatic ecosystems (Pauly and Christensen 1995). This is consistent with the usual 10 percent efficiency assumed over the last several decades, dating back at least to the work of Ryther (1969). This transfer efficiency should result in a tenfold increase in the available biomass for harvest at the next lower trophic level.

Few studies examine the relationship between fishing and ocean ecosystem productivity, but, in theory, an upper limit exists to the amount of production that can be removed by fisheries. Pauly and Christensen (1995) report that the primary production required to sustain reported catches plus discarded bycatch amounted to 8 percent of global aquatic primary production; however, this study includes freshwater fisheries as well. By ecosystem type, the requirements were only 2 percent for open-ocean systems, but ranged from 24 to 35 percent in freshwater, upwelling, and shelf systems (Pauly and Christensen 1995). Christensen (1998) further argues that only about 33 percent of primary productivity can be expected to be used for fisheries, meaning that for some areas, no growth in overall fisheries production can be expected.

However, if more productivity is available at lower trophic levels, one could argue that intentionally fishing down the higher trophic levels would be in the best interest of increasing long-term productivity from the ecosystem (Sprague and Arnold 1972). However, this may result in the harvest of species that were not previously viewed as desirable or those of lower economic value, and it might also require improved technology to catch and transform these new species into acceptable products (Garcia and Grainger 2005). The alternative scenario in which depletion of a top predator results in increased yields from valuable species is also possible. For example, shrimp and crab populations in the Northwest Atlantic increased greatly following overfishing of cod stocks (Worm and Myers 2003), and with high shellfish prices, the total value of the fishery is now greater than before the cod collapse.

A trophic index, called the FIB (fishing-in-balance) index, was developed by Pauly et al. (2000) to help judge whether additional yields expected by fishing lower in the trophic structure were actually occurring. The index is negative and decreases when the yield benefits are not realized by fishing lower trophic levels. In the Northwest Atlantic, this index initially increased as trophic levels and catches both increased and then declined after about 1970 and went strongly negative after 1990 without any yield benefits from fishing lower trophic levels. In a contrasting example for global tuna and billfishes as a group, while mean trophic level declined from 1950 to 2000, the FIB index steadily increased, suggesting the possibility of some yield benefits owing to fishing at lower trophic levels. Pauly and Palomares (2005) suggest that this resulted from a steady geographic extension of the fishery, but this is inconsistent with the fact that the mean trophic level was steadily declining during these years of increasing catches.

Are lower yields of the high predators more desirable in terms of consumption and/or profits than higher yields of lower trophic species? And what are the potential effects of manipulating fishery productivity on the overall ecosystem productivity? Certainly these issues demand more examination and the interactions between fisheries yield, ecosystem productivity, species composition, and fisheries market demands will need to be better understood and managed. Answering these types of questions will depend on the fishery and region in question, but the decision-making framework discussed in Chapter 4 may allow managers and stakeholders to consider these food-web interactions more explicitly in management decisions.

## RESPONDING TO REGIME SHIFTS

If ecosystem effects of fishing are to be identified, and perhaps more importantly, if ecosystem-based management is to be possible, the effects of decadal-scale climate variability on the inherent productivity of ecosystems and the fish stocks they contain will require explicit recognition and consideration. Toward this end, the effects of climate regime shifts on fish population dynamics have received increasing attention in recent decades. A climate regime can be defined as a persistent state in climate, ocean, and biological systems, with a regime shift being an abrupt, nonrandom change from one state or baseline to another (Beamish et al. 2004). Polovina (2005) reports that changes in biological baselines in response to regime shifts exhibit several characteristics. These are evident when abundances of various species covering a range of trophic levels that persist around long-term baselines suddenly and coherently shift to a new baseline level where they again persist. These shifts occur both for exploited and unexploited species. Additionally, shifts in ecosystems spanning broad spatial scales, including ocean basins, are temporally coherent (Polovina 2005). Interannual variation can and does occur within a regime, but the climate conditions indicative of the regime are relatively consistent and persistent compared to the magnitude of change that occurs between regimes (King 2005).

In the Pacific Ocean, and more recently in the Atlantic Ocean, decadal-scale climate regimes have been documented (Hurrell and van Loon 1997, Miller and Schneider 2000). These regimes are driven largely by changes in the relative position and strength of dominant atmospheric high and low pressure systems and are identified as the Pacific Decadal Oscillation (PDO) and the North Atlantic Oscillation (NAO), respectively. Scientific research has focused on both descriptions of how physical and environmental conditions vary among regimes and how these conditions propagate through food webs (Benson and Trites 2002).

The fisheries and ecological literature provides many examples of environmentally induced shifts in ecological communities large enough to be considered a regime shift. As Barange et al. (2004) point out, historical, fishery-independent records show large natural cycles of exploited fish populations in the absence of industrial fishing. For example, paleorecords of fish scales in anoxic sediments suggest large natural fluctuations in anchovy abundance off California during the last 1,700 years (Baumgartner et al. 1992) and in sockeye salmon abundance in the North Pacific over the last 2,200 years (Finney et al. 2002).

Changes in the population dynamics of fish species have been observed in phase with oceanic climate regime shifts across geographic regions (Francis et al. 1998, Benson and Trites 2002, Beamish et al. 2004). Kawasaki (1992) finds simultaneous variation in annual catches of Japanese, Californian, and Chilean sardine populations between 1910 and 1990. Strengths of year classes in several groundfish species, ranging from the California Current to the Bering Sea, show a dependence on the temperature regime; warm conditions were necessary but not sufficient for recruitment of a strong year class (Hollowed and Wooster 1992, 1995). Beamish (1995) reports other examples of fish population responses to climate regime shifts in the North Pacific.

Biological responses to regime shifts, especially in upper trophic levels, can lag or be masked by other processes and events (Miller and Schneider 2000). While direct, readily detectable responses to a regime shift have been observed in many cases, indirect food-web responses, variability in life history strategies, and longevity of some fish species may obscure or delay the onset of the effects of a regime shift on other fish populations (Benson and Trites 2002, Beamish et al. 2004). In different geographic areas, regime shifts could have opposite effects on the same species and, within an area, different species could respond in opposite ways (Benson and Trites 2002, Polovina 2005).

An interesting question is whether fishing, free from physical forcing, induces regime shifts. The discussion on trophic cascades above provides clear examples of ecosystem shifts induced by fishing at the top of the food web. This includes the northwest Atlantic example in Box 2.1, where the cascade contributed to an alternative dominant state, with a switch from groundfish to invertebrates. Fishing has clearly played a role in a number of radical shifts in marine ecosystems, but whether these changes constitute a regime shift is only determined by the longevity and the resiliency of fished species. Confounding the issues further, it is often difficult to tie changes in community composition to a single cause (Breitburg and Reidel 2005). Yet, Mangel and Levin (2005) make the case that many regime shifts could be driven in large part by fisheries. In addition, Collie et al. (2004) provide evidence that a discontinuous regime shift happened for Georges Bank haddock in 1965. The most likely trigger for this shift was the high fishing mortality during this time (Fogarty and Murawski 1998), but they were unable to rule out possible environmental contributions.

Clearly, intensive fishing can have drastic and long-lived effects on fished ecosystems. However, it is the combination of the two—climate change and fishing—that can result in the most severe effects. The response of fishery management can be critical in these situations, especially in a transition from a more productive to a less productive ocean ecosystem. And while a shift is sometimes

only definable several years after it has occurred, it may be possible and d
to account for more regular oscillations in management decisions if a stock is
known to be heavily fished and also sensitive to climatic regimes.

## RECOVERY, STABILITY, AND MULTIPLE STABLE STATES

Stability and resilience are emergent traits of all ecosystems in the sense that they are properties of the integrated whole. They attain great importance for all, but especially for exploited or managed systems if a tipping point or threshold is exceeded under perturbation, and the system collapses to a different structure and species composition that may be an alternative stable state. The stressors can be biotic or abiotic, intrinsic or extrinsic, and can vary greatly in magnitude and impact. Holling (1973) defines "stability" as the rate of return to some prior state after a perturbation. More stable systems would recover more rapidly with reduced fluctuations. "Resilience," on the other hand, depends on the maintenance of important relationships, thereby measuring the system's capacity to absorb stresses imposed on its original organization. To the extent that resilience implies ecological continuity or persistence and stability implies the capacity to recover from a perturbation, both terms are useful and commonly employed in the basic ecological and fisheries literature. However, managers must also be concerned about abrupt system collapses and the formation of an alternative state, a feature anticipated in the behavior of nonlinear systems and one with potentially severe ecological consequences. The primary issue is whether these states are stable and resilient.

The reality of and environmental threats to anthropogenically altered ecosystems are extensively discussed in the literature (e.g., Sheffer et al. 2001, Beisner et al. 2003, Carpenter 2003). The crux of the matter is the rate of return—if possible—to prior conditions. The sudden die-off of a Caribbean grazer, the urchin *Diadema* (Lessios 1988, Knowlton 1992), permitted benthic algae to smother the usual extensive coral cover (Hughes 1994). However, *Diadema* appears to be slowly recovering (Edmunds and Carpenter 2001), and it will be interesting to see if the ecosystem is able to switch back to its previous state. Plagues of carnivorous starfish continue to devour the Australian Great Barrier Reef. Anthropogenic influences, including fisheries, have driven the plague (Birkeland 1997); reduction of these forces could permit reef recovery.

Once a stock has been depleted by fishing, complex competition and predator-prey interactions may prevent reversal after fishing has ceased. For example, one hypothesis proposed to explain the lack of recovery of Newfoundland cod involves increased predation on cod juveniles by a predator that, prior to the cod collapse, was kept under control by cod predation (Walters and Kitchell 2001). Changes in fisheries and desired harvest species can also impede recovery. For example, off of northwest Africa, large resources of sea breams were intensely exploited in the 1960s and 1970s. After their collapse, they were progressively replaced by octopus stocks, and large fisheries for cephalopods developed as a consequence of increased market demand. However, sea breams were still taken as bycatch of the cephalopod fishery and the population never rebuilt to earlier levels despite not being targeted (Gulland and Garcia 1984).

The early debate about what was, or was not, a legitimate alternative and stable state has been extensively discussed in the primary ecological literature (e.g., Connell and Sousa 1983, Peterson 1984, Sutherland 1990). The debate identified two essential hallmarks: For community states to be comparable before and after perturbation, a common physical space must be involved, and the stressor must have vanished. That is, comparison of similar assemblages existing at spatially separated sites is inadmissible evidence, and the stressing force (e.g., persistent pollution or overfishing) must have been reduced to preperturbation intensity. The issue then becomes does the new state persist and, if so, for how long? Arbitrary judgments based on persistence, rate and degree of recovery, and magnitude of disruption are all required for analysis. For exploited populations, these are at the core of management decisions.

One can argue that anthropogenically influenced marine states are alternatives at both the population and community level. Whether they are stable may be difficult to distinguish in practice because a hysteresis from a single equilibrium value with a very long response time may appear stable. A single-equilibrium model should respond to a small intervention, albeit slowly. In contrast the multi-equilibrium model may not respond at all to a small intervention and may require a large intervention to initiate recovery.

Differences in recovery times of some stocks relative to others are not unexpected, however, and may be attributable in part to differences in the magnitude of compensatory reserve among interacting species. This fact alone could have a great deal of bearing on the way an ecosystem responds to the application or reduction of fishing pressure. In a relevant review, Rose et al. (2001) infer that life history and compensatory reserve are correlated and report that most of the world's fisheries target periodic life history strategists—fish that are relatively long-lived, highly fecund, broadcast spawners. Periodic strategists depend upon occasional large year-classes to persist over time scales relevant to the life expectancy of populations rather than that of individuals. Therefore, these species often are resistant to overfishing (i.e., have high compensatory reserve) because effects of removals are distributed over many year classes. But they generally do not recovery quickly (i.e., they are not resilient), even after fishing ceases, because of long population generation times.

On the other hand, species like the early maturing herring are opportunistic strategists (Winemiller and Rose 1992), that is, they are short-lived, colonizing type species that often exhibit boom-bust cycles in the face of fishing. These species often can and do respond quickly to changes in exploitation. In special cases, population levels can become so low that depensatory mechanisms may preclude population recovery no matter what curative measures are taken to restore the population (Shelton and Healey 1999).

From an ecosystem perspective, the mix of life-history strategies within communities that comprise the system and, hence, the mix in capacity of strong interactors to compensate for losses due to directed fishing or bycatch, will undoubtedly govern recovery times in ways that will be nonlinear and hard to predict. The economics of directed fishing usually deter fishing a stock to the point where depensation becomes a real danger. But for many of the most dramatic stock collapses, high economic value actually increases this risk, particularly for species that command extremely high prices at the market, such as large tuna or several invertebrate species. Bycatch species are also at risk since they are poorly monitored and assessed. At the very least, compensatory responses may be more difficult to assimilate and interpret from the ecosystem perspective, even though they have been the tenet of sustainable fishing for much of the history of fisheries science.

Key to this report is whether fishery management approaches can promote recovery from a fishery-induced regime shift. In the example presented in Box 2.2, the ecosystem did move from one state to another and back again when fishing pressures changed. Most of these changes were neither anticipated nor planned, but they demonstrate that shifting from one state to another is possible when fishing practices are changed. One could argue that these switches were not managed and that the favored state had not been established as a goal of management in a broad sense. Another example is the previously mentioned switch from cod to invertebrates in the northeast U.S. fishery. This shift was not planned and has not switched back even though fishing pressure has been reduced. But this unintended change can be seen by some as a positive outcome due to the high value of the subsequent shrimp fishery and therefore recovery to the previous state is not necessarily desirable.

If recovery of populations and food webs to earlier states is determined to be desirable, it will usually be a slow process, especially for fishes with older ages of maturation. In some cases recovery may not occur. Another lesson from the Hutchings (2000) study is that prevention of declines and collapses may be easier than designing for recovery. A prudent approach might be to prevent additional declines and collapses by more conservative population management, to develop and test alternative scenarios for food-web and community recovery, and to initiate long-term programs for recovery to serve as adaptive management experiments in the real world. Further, sustainability and the avoidance of persistent and unwanted alternatives critically depend on better data on the abiotic and biotic factors increasingly impacting all natural resources.

## MAJOR FINDINGS AND CONCLUSIONS FOR CHAPTER 2

**Ecosystem-level effects of fishing are evident.** Due to their extractive nature, fishing changes fish populations and therefore ecosystems. Reduced biomass and abundance, changes in size structure, genetic changes, and changes in trophic structure of ecosystems (including alternative states and altered productivity, species interactions, trophic cascades, and food webs) have all been documented as consequences of fishing. While some changes are the intended outcome of management actions, for many cases the measured effects are quantitatively and qualitatively more severe than management intended.

**Large declines in overall abundance of many stocks are evident even though the scientific debate continues regarding the magnitudes and implications of the declines.** Differences in reported declines of the same stocks have resulted from different assumptions, methods, and data used for the analyses; these issues are being worked out in the scientific literature. However, few deny there have been declines in numerous stocks and, in many cases, these declines have resulted in the loss of sustainable yields of particular species and species groups. Furthermore, the magnitudes of these declines vary spatially. Spatial variation is considerable in the condition of fisheries, different species, and geographically distinct stocks.

**Effects of fisheries removals can cascade through marine ecosystems.** Direct predator-prey interactions and indirect effects with lower or higher trophic levels are certain. Such cascading effects are often unpredictable and management actions frequently have surprising results that add an element of risk in predicting management outcomes. Further, some of the greatest long-term impacts of fishing have been observed in non-targeted ecosystem components. Many species, including marine mammals, seabirds, sea turtles, sharks, corals, kelps, and sea grasses have been affected negatively by fisheries either directly through bycatch or habitat damage, or indirectly through altered food-web interactions.

**Both fishing down the food web (sequential depletion) and fishing through the food web (sequential addition) are occurring.** Several lines of evidence suggest that the trophic structure of marine ecosystems is changing. However, studies based on landings data alone can be misleading because these data do not directly translate into accurate assessment of abundances in nature and may not reveal the underlying causes of trophic change. A single index of the mean trophic level of landings can be used as an indicator that one of these effects may be occurring. But because the management response will differ based on whether the cause is fishing down or fishing through, management decisions cannot be made from the index alone.

**Regime shifts caused by physical forcing, fishing, or both do occur.** Regime shifts caused by physical forcing are important to recognize for management purposes, especially when the shift results in lower ecosystem productivity. In addition, for systems that have changed primarily due to fishing activities, reversing ecosystem effects is not guaranteed because multiple stable states in communities may resist returns to an earlier state. The combination of the two creates conditions where recovery is even more uncertain.

**The shifting baselines phenomenon does occur.** When management actions and goals are likely to be influenced by shifting baselines, possible earlier conditions become a significant consideration. The policy implications are important with respect to setting rebuilding targets and strategies. For example, potential rebuilding targets for stocks may be set lower than what actually could be achieved for stocks released from fishing pressure.

**Realizing that there is a theoretical limit to the productivity that can be taken from the oceans and that we may currently be at or approaching that limit, food-web interactions will become increasingly important in future fisheries management decisions.** With the inability to change one ecosystem component without affecting numerous others, society will need to determine which ecosystem components are the most desirable for harvest, and then managers will need to implement policies designed to maximize this desired production while recognizing that this will affect other species. Future management advice should include ways to examine such interactions if the desired mix of fisheries and other ecosystem services are to be realized and achieved and if undesired or irreversible states are to be avoided.

# 3

# Considering the Management Implications

From a policy perspective, the ecosystem-level effects of fishing are of concern for two reasons: (1) the risk to overall long-term productivity of important commodities (i.e., harvested species) that can be obtained from the system, and (2) the need to maintain other ecosystem services in addition to these commodities. Maintaining productivity of commercially valuable species is linked to maintaining a functioning ecosystem that provides a range of additional services. These include climate regulating services and disease control, supporting services such as primary and secondary production, and cultural or aesthetic services (Millennium Assessment 2003). As the principal fisheries statute in the United States, the Sustainable Fisheries Act (P.L. 104-297) (adopted by Congress in 1996 as an amendment to the Magnuson-Stevens Fisheries Conservation and Management Act [MSFCMA] [P.L. 94-265]) includes a mandate to "protect the marine ecosystem," acknowledging the importance of ecosystem components and services beyond just fishery yields.

To move beyond managing for individual fishery yields, policies must be developed to fit within a framework of ecosystem-based management—considering the fishing effects of food-web interactions, bycatch, and habitat. Because all organisms are linked within a system, management strategies will have to make explicit tradeoffs among fished stocks, whether it is setting harvest rates, rebuilding strategies, or promoting other uses. A few examples of the management implications of trophic interactions and food-web effects, possible genetic changes, and physically induced regime shifts are provided here. The importance of developing multi-species harvesting strategies is considered, as well as some of the management structures that could be used to implement techniques to account for the ecosystem effects discussed in Chapter 2. This chapter focuses primarily on tradeoffs within and between fisheries; however, a larger suite of issues must be examined when considering tradeoffs in an ecosystem context. These issues are addressed in Chapter 4.

## FISHERIES MANAGEMENT IMPLICATIONS OF ECOSYSTEM INTERACTIONS

The consideration of multi-species interactions requires making decisions that involve explicit allocation tradeoffs both between food-web components and between user groups. In choosing harvest strategies in a multi-species system, it is impossible to avoid making de facto distribution decisions. Making these kinds of tradeoffs goes well beyond just deciding allowable catches for target species; bycatch must also be considered in most fisheries. Tradeoffs must also be made among fisheries, other commercial uses, and nonconsumptive uses of marine resources. These decisions are not scientific but instead decide the allocation of resources, although science still has an important role to play in informing such decisions (discussed further in Chapter 4).

Harvest strategies used in the United States involve the specification of biological reference points to determine target harvest rates as well as limits of fishing mortality and biomass that ought to be avoided. In this context, a natural first step is to determine how these harvest-rate targets and limits will be determined to account for species interactions. Should lower harvest targets be used for forage fish to protect the productivity of top predators? Should rebuilding targets be set taking into account that reestablishing depleted predator populations may impact fisheries that harvest their prey? Biological reference points depend upon life-history parameters, perhaps most significantly predation mortality (Collie and Gislason 2001). The effects of interactions can only be ignored if buffering mechanisms (e.g., predators switch between alternative food sources and prey have limited vulnerability to their predators) keep natural mortality rates from varying in response to changes in trophic structure.

This chapter presents examples of different fisheries scenarios and associated management implications that come into play when accounting for the food-web effects of fishing in marine ecosystems. These examples are simplified. They are intended only to highlight considerations that managers will begin to face as ecosystem interactions are incorporated into fisheries management and tradeoffs are explicitly made among species and users.

### Fishing a Single Trophic Level

The management implications of accounting for ecosystem effects of fishing vary depending upon the nature of the fisheries in question. The simplest setting (uncommon in practice) would be one in which fisheries only target the top trophic level. The major management concern would be to prevent the overexploitation of the top predator, accounting for any detrimental consequences of cascading effects to lower trophic levels. To achieve this, a modified single-species approach could be applied. In theory, an initial harvest rate would be selected that would sustain the population and prevent overfishing. This harvest rate would then need to be adjusted to account for the targeted species' role as a predator in the system. A harvest reduction may be necessary to diminish cascading effects via alternating increases and decreases of abundance for species down the food web. The magnitude of these adjustments would rely on knowledge of predator-prey interactions through the food web, the strength of these interactions, and appropriate food-web models to represent the dynamics.

Another single-species scenario (again uncommon in practice) occurs when a fishery targets a single intermediate trophic level. Similar concerns would exist for managers, but cascading effects would have to be evaluated both to lower and higher trophic levels. Setting a harvest target for these fisheries would have to incorporate information about the target's predators and prey, but in a bidirectional structure. If the population of the target species is reduced, then its predators would either consume less or would switch to preying on other species. These changes might be difficult to predict and hence generate unexpected results. For example, when great whales were hunted to low abundance in the North Pacific, it is hypothesized that orcas (the top predator in the system) switched from feeding on the whales to other marine mammals (Estes et al. 1998). As discussed in Box 2.2, orca predator switching may be the cause of a new trophic cascade that had effects on the entire ecosystem.

Setting the harvesting strategy for such a connected system would require the total allowable harvest to be divided between humans and the predators of the target species with total takes low enough that the population of the target species does not collapse. While these principles are straightforward in the abstract, in practice setting "single-species" harvest targets that take these ecosystem effects into account will be complex. As species abundances change, their interactions and the strength of these interactions will change as well. By necessity, the management process will need to be a flexible one that includes a monitoring scheme to provide feedback about the system status and an implementation strategy capable of responding to this feedback (Sainsbury et al. 2000). An iterative process would allow for the continuous monitoring and assessment of conditions and the incorporation of new knowledge about complex food-web interactions.

## Fishing Predator and Prey

A more common management scenario is one in which managers must set harvest regulations for multiple species that are harvested by different groups and that have direct predator-prey relationships. In this scenario, setting harvest targets for the species or complex caught by one group of fishermen will affect the potential yield of species targeted by other fishermen. Thus, managing for ecosystem effects will inevitably be more politically controversial because adjustments made to account for these effects will have allocation consequences to users. As an example, consider the strong predator-prey relationship between cod and capelin where each species is caught by different groups. Here, the desire to incorporate ecological interactions in decisions would need to involve explicitly acknowledged tradeoffs between user groups. For example, allowing more cod biomass for cod stock safety would require a reduction in allowable harvest for capelin over and above the reductions needed for capelin stock safety. Or, from the other perspective, constituents for the cod fishery would lobby for strong stock safety margins for capelin stocks, because that would allow greater sustainable harvests of cod. Capelin constituents would argue the opposite position, desiring higher catches of capelin.

Similarly, recognizing species interactions for rebuilding plans introduces an extra layer of explicit tradeoffs. If a cod collapse has allowed increases of another species, then a cod recovery plan will be seen as imposing costs on constituents targeting the newly abundant species. For example, accounting for interaction between cod and lobster in the East Coast system inevitably means that lobster constituents will see losses associated with cod rebuilding, namely the long-term reduction in lobster.

In a single-species world, the main tasks of reaching consensus revolve around convincing a specific fishery to take actions that are in its long-term interest, perhaps with short-term costs. But in a multi-species world, tradeoffs will have to be made, giving more yield to one fishery at the expense of another as a consequence of accounting for species interconnections.

### Fishing Multiple Trophic Levels

Managing fisheries that target multiple levels of the food web presents a more complex challenge than fishing one pair of predator and prey species. Adding lower trophic levels to the catch means that both predators and groups of prey are targeted. For example, in the Pacific Ocean, longline fishing reduced the abundance of the very largest predators; purse seines were then added to the fishery and caught a wider range of sizes, some of them juveniles of the apex predators and some the prey of the apex predators. In the multiple-trophic-level scenario, the reduction in the largest predators would reduce predation pressure, and therefore mortality rates, on the smaller fish. This would result in increased abundance of the smaller species or higher survival of juveniles, such that greater catches may be sustained than if the apex predators were not being fished. Fishing both groups could allow sustainable catches of apex predators while lower trophic levels are added to the catches. At the same time, negative effects on system biodiversity might be reduced because the fisheries are not merely cropping off the top predators through serial depletion. Overall, determining sustainable exploitation rates becomes a complex exercise in weighing tradeoffs in the catches from different trophic levels.

As discussed previously, exploitation of lower trophic levels creates competition between humans and apex predators for a common prey, but one in which the reduction in apex predators may allow greater catches of the prey species by the fishery than if only the prey species were harvested. But, reductions in fishing pressure on some of the larger predators would increase predation on the lower trophic levels and would require a reduction in fishing effort on these components of the food web. Again, a complex of policy conflicts and tradeoffs would arise from simultaneously harvesting apex predators and their forage base.

Management approaches for fishing several interacting trophic levels at the same time would need to be designed to account for impacts of fishing on the interacting system. The fishery literature has made explorations into fishing both predator and prey (Cox et al. 2002a, 2002b; Essington 2004; Essington and Hansson 2004; Hinke et al. 2004), but management protocols for dealing with this level of complexity are not in place.

## MANAGEMENT IMPLICATIONS ASIDE FROM TROPHIC INTERACTIONS AND TRADEOFFS

Some of the impacts to marine ecosystems discussed in Chapter 2 are not directly related to predator-prey interactions as presented in the previous examples. However, these impacts are equally important. They affect stock and ecosystem productivity and may ultimately result in lower or higher yields and must be managed accordingly.

### Preventing Genetic Changes

Recent evidence suggests that size-selective fisheries can alter life histories of marine fishes both phenotypically (Berkeley et al. 2004) and genetically (Conover and Munch 2002, Heino et al. 2002, Law and Stokes 2005). Even nonselective fisheries can reduce the reproductive lifespan of fishes that spawn over a number of years, thereby reducing their evolved buffer against uncertainty in larval survival (Heppell et al. 2005).

Fisheries based on a given species that target the largest fishes will tend to take the most rapidly growing fishes, resulting in selective pressure against fast growth. This may affect the genetic capacity for growth and ultimately production. In addition, recent studies with long-lived Pacific rockfishes have demonstrated that older females, in better physiological condition, produce larvae that are far more likely to survive than those produced by younger females (Berkeley et al.

2004). If the larvae of older females are important to producing robust year-classes, it may be especially important to maintain the older age groups in the population.

One approach to maintaining age structure would be to target the intermediate-sized fishes and not the largest fishes as is done currently. This could be accomplished using a slot limit where small fish are avoided or released to prevent growth overfishing and large fish are avoided or released to protect the most fecund spawners. However, slot limits can only work where gear or fishing methods are able to catch different sizes of fish (e.g., with hook size regulations) and where discard mortality is insignificant. But solely changing the sizes targeted would not be enough to increase abundances of older age groups; a reduction in mortality is still needed. In the case of Pacific rockfishes (*Sebastes* spp.) and other deep reef fishes (Coleman et al. 1999), slot limits seem unlikely to work. Another possibility is to protect whole communities by implementing marine reserves. For species with demonstrated maternal effects where maintenance of an unexploited age structure is critical, marine reserves may be the most reasonable alternative. Even for species that are moving in and out of the reserve area, protection of areas where the catchability of large fish is high will be beneficial. Further, in the case of highly migratory marine animals including fishes, it may be possible to provide protection using marine protected areas with moving boundaries (Hyrenbach et al. 2000, Norse et al. 2005). If fish, seabirds, marine mammals, or sea turtles, for example, migrate along specific pathways, it might be possible to reduce overall mortality by protecting them when they are passing through or aggregating in a particular area. Satellite telemetry coupled with new techniques in marine spatial analysis can provide models for these movements that link to oceanographic variables.

Fish such as Atlantic cod have also displayed much greater complexity in spatial success of reproduction than the continuous stock assumption would predict (Hutchings 2000, Hutchings and Reynolds 2004). This suggests that the location of fishing effort needs to be managed in conjunction with the total amount of effort to reduce genetic impacts. Spatially distributed fish stocks may be genetically distinct; in this situation, serial depletion of stocks not only reduces overall stock size, but also reduces the genetic diversity of the overall population by potentially eliminating locally adapted subpopulations. This process is well documented for Pacific salmon (NRC 1996b). Subpopulations have suites of adaptations that increase their fitness for the particular spawning location to which they home. When the fishery captures fishes from low productivity and high productivity subpopulations that mix on the fishing grounds, substantial depletion and extinction of the so-called "weak" stocks are probable. This is a serious problem with migratory anadromous fishes like Pacific salmon.

## Responding to Regime Shifts

Two different forces can induce a shift in the marine ecosystem. The first is by fishing pressure alone. A fisheries-induced regime shift drives the system into an alternative state with different relative abundances and a different structure than the previous system. The degree to which the ecosystem services provided under the new state will be inferior to the old one is largely uncertain. However, the shift may be irreversible and therefore a precautionary approach would prescribe taking measures to avoid fishing at high harvest rates capable of inducing a regime shift. The second cause of regime shifts are large-scale climatic drivers such as the Pacific Decadal Oscillation or the North Atlantic Oscillation, which cannot be avoided. Fishery management implications arise because fishing may amplify the effects of climate changes. How should managers anticipate and adjust fishing pressure during a climatically driven shift, especially a shift from high to low productivity?

In periods of high productivity, the fishery would be expected to expand, and it should do so if managers choose to take advantage of the high production. During a shift to a period of low productivity, fishing effort would need to be reduced so that fished populations and communities were not depressed to unsustainable levels or to levels that would prevent or impede recovery when the climatic shift returned to more favorable conditions. But scientists and managers can only reasonably predict some climatic shifts, such as the Pacific Decadal Oscillation. Thus, unavoidable lags in fishery management responses for unforeseen shifts will potentially exacerbate ecosystem-level effects and thereby make the ecosystem even less productive. Such lags would be especially problematic if the ecological responses to physical changes are nonlinear and rapid as suggested in the analyses by Hsieh et al. (2005).

Polovina (2005) provides a summary of studies that have examined the issue of what constitutes optimum management strategies for fisheries that undergo regime shifts. Most of these studies use models to simulate low-frequency variation in survival and carrying capacity to represent climate-induced regime shifts (Walters and Parma 1996, Spencer 1997, DiNardo and Wetherall 1999, Peterman et al. 2000, MacCall 2002). Two different strategies emerge from these simulations. The first, which performed well under some circumstances, employs constant harvest rates set irrespective of environmental regimes (Parma 1990, Walters and Parma 1996, DiNardo and Wetherall 1999). In other simulations representing different types of systems, results favor a regime-specific harvest rate strategy over a constant harvest rate strategy (Spencer 1997, Peterman et al. 2000, MacCall 2002). To a large extent, adjusting harvest rates in response to changing environmental conditions will depend on the frequency of regime shifts relative to the life span of the target species. The benefits of regime-specific strategies will increase when the magnitude of potential change in productivity is large and when climate regimes persist for longer time periods.

## DEVELOPING MULTIPLE STOCK HARVEST STRATEGIES

In principle, the maximum sustainable yield (MSY) concept[1] and related management targets and strategies can be extended to multiple stocks as a basis for policy, although setting multiple species harvest targets is an inherently more complicated task, as indicated in the previous examples. The important conceptual point is that if species are linked through trophic, food-web, and habitat interactions, accounting for those linkages inevitably means considering harvesting strategies for all species simultaneously, in a manner that recognizes the interconnections. This in turn means that multi-species harvest strategies for any species cannot be determined independently of those for other linked species. This presents important scientific questions as well as questions relating to what the systemwide objectives should be.

Multi-species harvest targets can be established only if the goals for the systems are specified. One possible management option is to use the multi-species analogue of the single-species MSY concept. Under this objective, harvest targets or reference points that achieve the system-wide sum of sustainable physical yield could be computed (at least in principle). The suite of harvest targets that maximize the sum of systemwide yield would differ from their single-species analogues in ways that depended upon trophic interactions (e.g., Beddington and May 1977, May et al. 1979). Another possible management objective might be to maximize the systemwide value-weighted sum of sustainable yields. Again, both trophic interactions and relative economic values of each species would interact in ways that determine the suite of multi-species reference points. Christensen and Walters (2004) show that maximizing value-weighted sustainable yield generally implies larger systemwide biomass levels than maximizing unweighted yields. They also show that the ecosystem configuration that maximizes a systemwide objective is not necessarily similar to what would emerge by maximizing independent species-specific target yields.

A comprehensively conducted multi-species analysis of harvest targets and reference limits could also, in principle, account for nonconsumptive uses. This could be done, for example, by incorporating information on the manner in which nonconsumptive services depend upon biomass size and characteristics. Species with high values for nonconsumptive services would compete with consumptive values in a comprehensive analysis, in some cases implying maintenance of high biomass levels and low (or zero) harvests. At present, these kinds of simulations are difficult to do because while a great deal is known about economic values associated with consumptive uses like fishing, comparatively little is known

---

[1]The maximum sustainable yield (MSY) concept is the concept that there is a maximum average catch or yield that can continuously be harvested from a stock under existing environmental conditions without affecting significantly the reproduction process of that stock. See Appendix D for a glossary.

about nonconsumptive values. Methods that can account for all these tradeoffs—consumptive uses, nonconsumptive uses, and ecosystem services—are available, but are in their infancy.

## The Role of MSY in Multi-Species Management

The policy guidelines available for managing single stocks of fish are quite well specified. The MSFCMA (P.L. 94-265) states that Congress finds, "Fisheries resources are finite but renewable. If placed under sound management before overfishing has caused irreversible effects, the fisheries can be conserved and maintained so as to provide optimum yields on a continuing basis." Furthermore, Congress stated, the term "optimum," with respect to the optimum yield from a fishery, means the amount of fish which:

(A) will provide the greatest overall benefit to the Nation, particularly with respect to food production and recreational opportunities, and taking into account the protection of marine ecosystems;

(B) is prescribed as such on the basis of the maximum sustainable yield (MSY) from the fishery, as reduced by any relevant economic, social and ecological factor; and

(C) in the case of an overfished fishery, provides for rebuilding to a level consistent with producing the maximum sustainable yield in such fishery.

The concept of sustainable yield results from the intrinsic ability of biological populations to compensate for increasing fishing pressure by increasing their productivity. Density-dependent compensation results in some surplus production that can be harvested on a continuing or sustainable basis. As a conceptual framework this is a straightforward proposition.

However, managing by species-focused MSY on targeted stocks does not take into account the food-web interactions discussed previously. So even if only a single stock is being managed, the harvest level must be set with these considerations in mind. In addition, in multi-species fisheries with bycatch of non-commercial stocks, ignoring bycatch might result in unacceptable depletion of species that are important for ecosystem health. Single-species-focused MSY management also neglects possible habitat damage. Yet it is important to note that current management measures have begun to incorporate these larger effects and have succeeded in reducing bycatch and protecting habitat.

Nonfishery concerns must be addressed as well. The opportunity cost associated with other ecosystem services increases with higher levels of biomass and other characteristics closer to pristine states. Additionally, when cascading effects are caused by the fisheries, even maintaining harvest rates at the single-species MSY level could mean that developing other uses is precluded. For example, if the abundance of a forage fish species is important to maintaining other ecosystem components, maximizing fishery yield neglects consideration of the role of that forage species in sustaining the dependent species and the ecosystem.

These types of considerations are explored in the literature from both theoretical (e.g., Beddington and May 1977, May et al. 1979, Clark 1985) and applied perspectives (e.g., Collie et al. 2003, Walters et al. 2005). Indications are that single-species MSY policies may set harvest rates too high when ecosystem interactions occur, especially when the target species are prey to other species whose productivity is to be preserved. For example, Walters et al. (2005) examine the performance of single-species harvest rate policies for 11 model ecosystems representing a wide range of different systems. They demonstrate that widespread application of single-species MSY fishing mortality rates ($F_{MSY}$) would in general cause severe deterioration in ecosystem structure, in particular the loss of top predator species. However, in their study $F_{MSY}$ was applied to all species in the ecosystem, including forage fish that have been only lightly fished in the past and that would provide alternative food sources to top predators when some of their prey are fished. For most of the ecosystem models, they conclude that "(1) yields under a many-species $F_{MSY}$ policy can diverge grossly from single-species predictions, and (2) the direction of divergence is not consistently related to trophic level" (Walters et al. 2005, p. 566). Better performance overall occurred when they simulated a more precautionary harvest rate policy in which the fishing mortality rate was set to 70 percent of the single-species $F_{MSY}$. However, this is only one example and clearly more such studies are needed to advise management regarding what deviations from single-species MSY would be necessary to maintain ecosystem integrity.

The following section lays out a framework for evaluating fisheries management strategies in an ecosystem context, but, in the short-term and for systems where this framework cannot be implemented, a precautionary approach (e.g., Restrepo et al. 1999; FAO 1996) should be applied, choosing some percentage of the MSY as the target harvest rate, as in the Walters et al. (2005) study. Considering that historically many fisheries have substantially and repeatedly exceeded the harvest rate that would ostensibly produce MSY, a margin of safety for application of MSY estimates is appropriate. At the very least, when MSY-based rules are applied in systems without accounting for species interactions, using $F_{MSY}$ as a limit reference point instead of as a target could be an essential step in guarding against future overfishing (Mace 2001). Such an approach is called for in the United Nations Treaty on Highly Migratory and Straddling Fish Stocks (United Nations [UN] 1995). But, it is essential to note that regions with harvesting strategies that were designed using model-based scenario analysis (as described in Chapter 4) would tend not to support using a fixed percentage of $F_{MSY}$ for all species as the preferred management action. In some cases, it may be deemed necessary to exceed $F_{MSY}$ to achieve larger, ecosystemwide goals. But such approaches will require sufficient data about the system to reasonably evaluate potential outcomes.

## MECHANISMS FOR IMPLEMENTING MULTI-SPECIES HARVESTING STRATEGIES

The model-based scenario approach discussed in Chapter 4 is the essential step for evaluating alternative fishing strategies that can address ecosystem concerns. However, it is first necessary to discuss some of the regulatory schemes that can be used to manage fisheries since these differing approaches will need to be tested in model simulations and weighed as potential options when deciding tradeoffs, both between competing fisheries and between fisheries and other uses.

Ecosystem and food-web considerations might be accounted for in fishery management by various mechanisms. Fisheries are primarily managed by direct or indirect controls on either inputs (e.g., effort, gear type and configuration, and time and area closures or openings) or outputs (e.g., catch in weight or numbers, limitations on landing certain sizes or species of fish, and limitations on bycatch amounts). Direct controls regulate the input explicitly; for example, a limited-entry program fixes the number of vessels, or season-length restrictions close the fishery upon attaining targeted catch or landings. Indirect controls are intended to limit inputs or outputs by constraining other features of the fishery, such as gear restrictions that affect the efficiency of fishing or area closures that prohibit fishing (Box 3.1). For all of these management tactics, ecosystem considerations could, in principle, be included in the determination of harvest strategies that address a broader set of impacts and conserve ecosystem structure and function.

Most fisheries management in the United States and internationally relies on output controls with catch quotas as a primary regulatory objective, accomplished by some input controls on gear, areas, and seasons. From an ecosystem perspective, addressing the manner in which input controls are chosen and used may be more important than the choice of output controls. This is because ecosystem effects often result from the specifics of how fishing effort is exerted, rather than the absolute level of removal of target species. Habitat impacts and bycatch, for example, result from the level and type of fishing effort, regardless of how much is landed. In effect, this means that fine-tuning fishing effort configurations is a critical mechanism for managing ecosystem effects. Output goals are some measure of the performance of input controls in this sense. If the fishing capacity, fishing time, gear, and areas allowed are properly set, then the output control, such as the amount landed, should serve only as a backstop against overfishing rather than as the primary control mechanism. In the next section, two conceptual alternatives are discussed for managing inputs and outputs in fisheries, namely top-down and bottom-up structures.

### Institutional Structures and the Regulation of Fishing Effort

Fishing effort can be managed for either single- or multiple-species objectives with two different institutional structures. By far the most common method

> **BOX 3.1**
> **The Use of Marine Protected Areas**
>
> One possible approach to address ecosystem concerns is the use of protected or reserved areas (NRC 1999a, 2001). The major benefit of "no-take" areas is the ability to protect both target stocks and bycatch species from harvest in cases where species have low mobility relative to the size of the reserve. The habitat needed for commensurate species is also protected, presuming the requisite enforcement occurs. Well-designed closed areas or marine protected areas can be important buffers against uncertainty in controlling fishery harvest rates (Stefansson and Rosenberg 2005). They can also protect against genetic changes in intense size-selective fisheries and in long-lived species with demonstrated maternal effects (Heino et al. 2002, Berkeley et al. 2004). Their benefits, however, will depend on sound management outside of the reserves to maintain fishing effort within ecologically sustainable limits (Hilborn et al. 2004), thereby preventing the simple redirection of fishing effort from one area to another. The use of reserves and protected areas, or of networks of reserves, is not a panacea, but it is one consideration—within a suite of management options—for mitigating ecosystem-level impacts of fishing, especially in areas that have been severely degraded.
>
> On an experimental level, strict "no-take" areas would be the only way to eliminate fishing pressure, if only for a set amount of time, to observe whether stock and ecosystem recovery is possible. If set aside for a long enough time, a marine reserve of the correct size could assist in resolving issues with shifting baselines, allowing scientists and managers to quantify how much the surrounding area has changed due to fishing.

utilized in both the United States and internationally is what is known as top-down control, meaning centralized determination and enforcement of total output via input controls, and an absence of secure individual-access privileges. Under top-down control systems, harvest target goals or limit points are set, and then input controls are chosen and implemented by some management body to achieve these goals. In U.S. fisheries, these output and input control decisions are vested mostly in the Regional Fishery Management Councils. A commonly used procedure is to set harvest targets or limits within a backdrop of MSY concepts, and then use seasonal closures once accumulated harvest reaches the target. Additional measures are also commonly added onto basic total effort controls on commercial fishermen to address a range of noncommercial fishery goals, including protecting species from excessive bycatch, incorporating mammal and bird protection regulations, and allowing other user services in addition to commercial fishing. For example, bycatch may be regulated by gear restrictions (e.g., requiring turtle excluder devices) or by closing the season for a target if a bycatch limit is reached.

In principle, existing top-down regulatory procedures can be adapted to account for ecosystem effects in a more explicit and less ad hoc fashion. This would involve two steps. The first would require the specification of a new set of systemwide harvest targets that account for trophic interactions as well as rules that limit ecosystem impacts such as habitat loss and loss of biodiversity. The second step would be to determine a new suite of regulatory actions to limit effort according to the modified harvest rules. With top-down regulations, Councils set harvest objectives and targets and then determine constraints on individual fishermen's decisions to attain the objectives. All fishermen and other user groups can play an indirect role in setting harvest rules by lobbying the Councils, but ultimately the final decisions are left to the top-most layers in the system, namely the Council membership and the Secretary of Commerce. Therefore the tradeoffs between fishing groups are made overtly by managers or by a process determined by the political system. Top-down approaches have been effective in reducing or preventing overfishing in many fisheries, although in other cases they have failed to effectively constrain effort and avoid overexploitation.

The most important drawback of top-down governance institutions is that they maintain an adversarial relationship between regulators and regulatees. That is, fishermen are seen by regulators as needing control and restraint, and hence their access to the resource is left tenuous and uncertain, and tightly controlled by regulations. By leaving resource access insecure, this system generates perverse individual incentives to increase fishing capacity, which must, in turn, be met by further imposition of controls by regulators. Top-down systems with insecure access privileges thus generate a "race to fish" which, if it does not actually subvert the intended control over the resource, nevertheless generates continual increases in capacity and economic waste. In most fisheries, the race to fish continues as long as growing markets keep increasing prices, leading to shorter and shorter seasons, unevenly applied effort, poor quality product, and wasteful investment in distorted fishing gear and capital, all wrapped up in an adversarial process between regulators and fishermen.

From an ecosystem perspective, recreational and commercial fishermen have little incentive in a top-down system to limit their ecosystem-level impacts other than through altruism. Because the race is on, ecosystem impacts take a back seat to the fundamental uncertainty in the process, which is securing a share of the harvest target before one's competitors do. In the race-to-fish setting, a spiral is then set into motion, with managers tasked to create ever more complex regulations to try to reduce the fishing effort.

An alternative to the top-down approach is to implement so called bottom-up management systems with secure access privileges. Bottom-up approaches still require that harvest rules be set in some fashion, but they eliminate the need to micromanage the details of effort and input decisions with regulations. The key to this alternative system is the creation and allocation of harvest-access privileges that eliminate the race-to-fish incentives that exist under top-down management.

In decentralized, bottom-up regulation, fishermen have secure access privileges to a fraction of the total allowable catch for each species. These may be individually denominated privileges, as with individual transferable quotas (ITQs, see NRC 1999b), or they may be group allocated privileges, such as to a harvester cooperative or a community. A few such systems exist in the United States in Alaska, and both domestic and international examples have resulted in important changes to fishermen's behaviors. Perhaps the most important lesson from the adoption of these systems is that, with secure access privileges, the incentives generated for individuals are radically different from those under top-down command and control with insecure access. With secure access privileges, whether granted to an individual or group, fishermen no longer need to race to fish because their allocation guarantees them access. In this environment, behavior switches dramatically from catch maximizing to value maximizing. Fishing is slower, fishing capital is (generally) downsized, redundant inputs are eliminated, and new innovations in the market are stimulated to increase value of harvest. On the cost side, these systems may require more enforcement and monitoring to prevent dumping (e.g., discarding of a portion of the catch to stay within allocations), information fouling (e.g., falsifying records such that the allocation is undermined), and cheating (e.g., landings outside of the allocation, illegal transfers to other allocation holders), and they always involve contentious initial allocation decisions regarding who is granted privileges and how much allocation is granted.

In bottom-up systems with secure access privileges, the tactical decisions are left to the fishermen about how to conduct their fishing operations to maximize the value of their allocations. While these systems have generated important positive impacts on commercially targeted species, they do not address all important ecosystem impacts of fishing. However, it is possible to modify the basic structure of the systems where access privileges are predicated on ensuring that impacts to other ecosystem components are minimized.

Bottom-up governance systems may also promote the development of fishing methods that more efficiently reduce ecosystem level effects. For example, if limiting noncommercially valuable bycatch is deemed necessary, allocation privileges for the target species can be extended so that fishermen also have allocations of bycatch determined using multi-species approaches to set reference points. In other words, managers could set limits on acceptable levels of noncommercial bycatch, allocate these as bycatch harvest privileges to individual fishermen (or groups), and allow fishermen to best choose methods to avoid using their allocations. While enforcement and accurate reporting remain an issue and are generally more costly under access privilege systems, there are important advantages to this system compared with conventional top-down systems. For example, a multi-species fishery might be closed by regulators when some reference level for bycatch is reached. Under a bottom-up system with bycatch allocations, fishermen "use up" their bycatch allocations as the level of bycatch deemed acceptable for ecosystem services is approached. Furthermore, if a bycatch allocation can be traded among fishermen, they take on their own market value during the season. Fishermen can then either use or sell their allocations in a tradable system; every ton of bycatch avoided presents an opportunity to sell a ton. This creates an automatic and continuing incentive to adjust fishing behavior to avoid bycatch. The same notion of trading and accumulating harvest privileges could be used to account for nonconsumptive services associated either with components of a system or even particular areas of marine ecosystems. But sustaining these decentralized incentive effects is not easy; enforcement and monitoring are important components of maintaining such a system since the incentives to cheat and underreport bycatch are similar to those in a system with directed-catch allocations.

An additional important effect of designating harvest-access privileges is that the privileges become securities, in the same sense that holding a share of stock promises access to a flow of future dividends. The importance of securing access privileges is that they generate a stewardship ethic that motivates concern about the long-term health and productivity of the system, and the privilege can and should be coupled with responsibilities for stewardship in order to maintain that access. Individual transferable quotas and membership values for co-ops thus take on values that are similar to farmers' land values. And with embedded values, owners are compelled to become stewards with long-term interests that preserve the values as well as the access if the privilege is tied to specific management needs (e.g., bycatch reduction, accurate reporting). Because of this embedded value, there is an additional incentive to make decisions that increase long-term values. This presents both opportunities and challenges for dealing with multispecies interactions in decentralized systems.

Furthermore, fishermen may be compelled to seek out quota rearrangements with other fishermen to optimize the value of their holdings. And "other fishermen" may include those with whom a particular group of rights holders interact via the interrelated nature of their target species. For example, in the earlier cod–capelin example, cod fishermen might find it desirable to purchase harvest access privileges held by capelin fishermen to account for the predator-prey ecosystem effects of having a larger biomass of capelin to support the cod. Thus the difficult political decisions that we described as being required to manage multi-species systems might be allowed to occur spontaneously under some circumstances.

The implications of bottom-up, access privilege-based systems are only just beginning to be understood as new examples are implemented in the United States and around the world. Hundreds of species are managed with these kinds of systems, which range from individual transferable (and nontransferable) systems to harvester cooperatives, regional-area-based cooperatives, and territorial-use right systems. The pros and cons of these have been debated for the past three decades; numerous summaries exist (see NRC 1999b). We will likely see more and not fewer of these institutions being adopted in the future. But, at this point, these discussions are simply illustrative of the possible consequences of using bottom-up approaches to achieve multi-species ecosystem objectives. Much more needs to be discussed and more research conducted on these issues. The questions that arise relate to whether decentralized and voluntary reallocations ought to be allowed or encouraged to achieve ecosystem-based fisheries management and, if so, what rules and institutions might facilitate them.

## OVERCOMING REGULATORY CONSTRAINTS TO SETTING MULTI-SPECIES REFERENCE POINTS

An overarching framework does not exist within the current U.S. management system that explicitly addresses ecosystem management of marine systems across the various sectors of human activity. Institutionally, management is organized largely with respect to sectors of human activity: fishing, coastal development, water quality, and so forth. Even with area-based authorities such as the National Marine Sanctuaries, most of the regulation of specific activities, such as fishing, is left to specialized agencies. This system is not conducive to determining goals and tradeoffs between sectors and between users.

Within multifunctional agencies such as the National Oceanic and Atmospheric Administration (NOAA) or the Environmental Protection Agency (EPA), little coordination exists between programs that manage activities affecting marine ecosystems (U.S. Commission on Ocean Policy 2004). In part, this lack of coordination stems from the statutory mandates currently in place. Individual agencies have mandated responsibilities that do not necessarily allow them to develop management actions that are more broadly based and coordinated across sectors of human activity. With respect to fisheries, while the MSFCMA calls for conservation of ecosystems on which fisheries depend, the national standards for management plans do not clearly call for coordination with other management actions outside the fishery sector. Nor is there a clear mandate in the national standards for managing the ecosystem effects of fishing, other than through consideration of fisheries habitat.

Other existing laws, which require that management actions focus on certain single species, confound the issues. The Marine Mammal Protection Act (MMPA) and the Endangered Species Act (ESA) generally regulate activities on a species-by-species basis for species already in crisis. However laudable these efforts may be, they too tend to ignore the interaction and interdependence of marine ecosystems in favor of regulation on a species-specific basis. Such narrowly focused regulation can have unintended effects when cross-linked with fisheries and can make management decisions more difficult. Consider the potential conflict between restoring Pacific Northwest salmon populations and the predation on salmon by sea lions protected under the MMPA (NOAA 2002).

Resolution may be even more difficult when both partners in a predator-prey interaction have some degree of federal protection. Sea otters are in precipitous decline in Alaskan waters, most likely due to increased predation by killer whales

(Estes et al. 1998, 2005). Both are protected under the MMPA. In California, sea otters can control populations of commercially and recreationally valuable invertebrates (e.g., clams, sea urchins, and abalone) but also are a plausible threat to the severely endangered white abalone. Both otters and abalone are ESA listed. Slight attention has been paid to these regulatory dilemmas, which are certain to increase as management's perspective embraces an increasing variety of linked species. As described in this chapter, consideration of ecosystem effects requires explicit consideration of tradeoffs in ecosystem services under different management actions. In effect, the current statutory structure precludes certain tradeoffs unless some overarching authority for ecosystem-based management is created.

## MAJOR FINDINGS AND CONCLUSIONS FOR CHAPTER 3

**Managing fisheries within an ecosystem context will require accounting for food-web interactions and trophic effects and making tradeoffs between species or among fisheries and other uses.** It is essential that tradeoffs be made among fisheries, other commercial uses, and nonconsumptive uses of marine resources since value and/or yield are unlikely to be maximized for all species. In an ecosystem context, the potential productivity or value of each resource depends on the management decisions made about other linked species. Accounting for species linkages will mean designing and implementing harvest strategies that recognize these interconnections.

**Single-species MSY policies are unlikely to be sufficient for future management because these measures do not take into account species interactions and food-web effects nor do they consider nonconsumptive ecosystem services.** Preliminary evidence indicates that $F_{MSY}$ policies can set harvest rates too high when food-web interactions occur. However, whether single-species MSY harvest policies lead to harvest rates that are too high or too low will depend on the particular species, its trophic interactions, and, ultimately, on management goals, in particular how tradeoffs between competing uses are resolved. If the impacts of alternative harvest rates have not been examined using interaction models, implementing harvest rates at some fraction of single-species $F_{MSY}$ is likely the best protection against immediate overfishing. At the very least, $F_{MSY}$ should be implemented as a limit reference point and not a target.

**A variety of new regulatory mechanisms and institutions ought to be considered to help implement ecosystem-based management approaches.** Successful accounting for ecosystem effects will require, at the first level, accounting for multi-species interactions. But it will also require mechanisms to deal with new and politically contentious allocation decisions within fisheries, and between fisheries and other nonconsumptive uses. Furthermore, there is a continuing need to consider governance structures that align fishermen's incentives with long-term stewardship. Bottom-up, access privilege-based systems may hold enhanced promise to address some of the new issues raised by ecosystem considerations.

**Existing laws and agency structures will need to be examined against a wider mandate to implement an ecosystem approach to management.** Several regulatory mandates and agency programs have been created specifically to protect certain species, and they are especially important for species in danger of extinction. However, such single-species-focused mandates and programs may result in conflicting goals in a multi-species or ecosystem approach to management. Dissolution of these single-species protections is not the answer; rather, management institutions must recognize that one protection may preclude the other. An overarching mandate that allows explicit consideration of tradeoffs is needed to resolve the difficulties of reconciling the existing mandates.

# 4

# Informing the Debate: Examining Options for Management and Stewardship

Setting policy goals for the conservation of ecosystem services requires going beyond simply managing for fisheries yield and the reversibility of fisheries-related depletion. The term ecosystem-based management has been used by the U.S. Commission on Ocean Policy (2004) and the Pew Oceans Commission (2003) to mean developing ecosystem-level goals that are multispecies focused and that consider multiple kinds of human activities that are tied to healthy marine ecosystems. This means that both consumptive and nonconsumptive uses are weighed when deciding management actions and regulations for the oceans and coasts as a whole.

However, even if the overarching policy goal for marine ecosystems is to manage both consumptive and nonconsumptive uses in an integrated approach, successfully carrying out the objective ultimately requires determining what the overall ecosystem goals should be. What level of productivity is desirable? How much risk of irreversible change is acceptable? What do we—as a community—want our ocean environments to be? These are all questions about social values.

One possible goal is to strive for pristine ecosystems that resemble natural conditions with no human impact. This is neither possible nor practical, for it would be impossible for humans not to have an impact on ocean ecosystems. Indeed humans are part of marine ecosystems. Even if fishing pressure were eliminated, there would still be impacts from coastal development, tourism, transportation, and many other uses. Thus, it must be decided what mix of these uses is most desirable based on measured or perceived benefits. These are not easy decisions. Adequate scientific knowledge from both the natural and social sciences is important for delineating options and illuminating choices, but science alone cannot address these issues. The path forward will require melding of the understanding of ecosystem processes, the human dimension, and the possible policy options. Tradeoffs between conflicting goals can then be decided with input from diverse users, including tradeoffs needed to implement restoration or rehabilitation activities when deemed necessary.

This chapter discusses the application of model-based scenario analysis for shaping fishery management goals and the current capabilities for conducting such analyses. Also presented is a discussion of environmental ethics and the public involvement needed to make sound decisions for our future and ensure that all uses of ocean resources are represented when setting fishery management policies.

## EVALUATING STRATEGIC MANAGEMENT OPTIONS

Fisheries management in the United States in recent years has tended to follow prescriptive policies defined in terms of nonspecific biological reference points used to set targets and limit harvest rates and to specify biomass thresholds to be avoided. In this management setting, the scientific support for management decisions is largely couched in terms of how different levels of catch or effort controls fare against the accepted generic standards. For example, annual quotas are estimated so as to meet a target exploitation fraction with a low probability of exceeding the thresholds. A stock assessment is conducted annually and the resulting estimate of stock size is normally multiplied by the target exploitation fraction to calculate the allowable catch. In terms of management goals, a de facto decision about the "best" policy is implied in the choice of the generic reference points. Thus, the role of the annual stock assessment is largely limited to informing tactical decisions, such as the choice of an annual catch quota, instead of being concerned with evaluating the consequences of different strategic policy choices for the ecosystem and for all different stakeholders. Ecosystem considerations are discussed in regular stock assessments, and environmental impact statements are required for major fishery management actions, but in general these actions do not involve a comprehensive evaluation of management strategies and there is no legal requirement to account for these ecosystem interactions.

Certainly science—including social and economic science—has a much larger role to play in informing strategic policy choices. In some countries and for other U.S. resource management organizations, fishing strategies or management procedures are designed by taking into account the nature of the system and the specific management issues involved. Different candidate strategies are tested using simulation models, and their performance is evaluated by examining graphical outputs of model projections and various performance statistics that measure policy outcomes according to different, often conflicting, management goals. The role of fishery scientists in these cases is to integrate all available information (scientific and empirical) to help assess the likely consequences of alternative policy choices. This is best done formally using simulation models to test the performance of candidate policies across a variety of scenarios/hypotheses that represent all possible futures in light of all available evidence. The challenge for scientists is to identify the series of scenarios that capture the uncertainty existing about the system dynamics and to assign each a relative plausibility.

A formal approach for the evaluation of alternative management strategies was pioneered at the International Whaling Commission (IWC 2005) and has been mostly applied to the design of catch-control rules for industrial fisheries (e.g., Butterworth and Punt 1999, Parma 2002a, Smith et al. 1999). The approach involves: (1) formal specification of scenarios used to represent the dynamics of the exploited system and their relative plausibility, (2) identification of candidate policies considered feasible a priori, and (3) identification of graphical output and definition of a series of performance statistics used to measure policy outcomes that reflect the interests of all stakeholders.

The first component is strictly a scientific endeavor. Assessments are conducted based on various competing assumptions about key processes in the dynamics of the stock and its fishery, and the results of these assessments provide the basis for identifying alternative scenarios to be used for policy evaluation. However, the role of the assessment is not to come up with the best estimate of stock size, but to "condition" the simulation models to the available historic information. The different scenarios may represent a range of productivity, maximum stock size and depletion levels, and/or different structural models of the system dynamics, as well as different relationships between the fishery indicators used for monitoring the underlying stock dynamics.

Management strategies then go well beyond the specification of reference points. They not only specify the feedback rules used to calculate input or output controls, but also specify the data needed to implement those rules. In this way, the evaluation of management strategies places due emphasis on the information gaps and uncertainties specific to the fishery in question, as well as on the problems that need to be addressed for the implementation of the different candidate policies. The results of the policy evaluation can be summarized in the form of a decision table that provides the outcomes of different candidate policies across a range of model scenarios. Some policies tend to be more robust to the uncertainties than others, but ultimately some fundamental tradeoffs between conflicting management objectives need to be made.

The goal of the analytical exercise is not to build models that are able to predict what will happen, but to build a series of models and hypotheses of what *may* happen and to assign relative plausibilities to them so that tradeoffs between conflicting management objectives will be explicit when decisions are made. These tradeoffs must be decided not only between competing fisheries, but also between fisheries and other uses of marine resources.

The last two components of the approach described involve societal values and choices, and therefore require input from all parties so that the different perspectives are reflected in the selection of candidate policies. (Greater stakeholder participation is discussed in a later section.) Although experience with formal evaluation of fisheries management strategies around the world has mostly focused on single-species management as described above, the design framework is flexible and could be applied to assess the consequences of alternative policies in multi-species systems (Sainsbury et al. 2000).

## Using Ecosystem Models for Policy Screening

The strength of single-species policy analysis is its strong reliance on past data to develop the model scenarios. Conditioning is critical because by fitting the models to historical data, the universe of plausible models is constrained to those that can reproduce historical trends. Conditioning of single-species models is supported by significant development of statistical techniques and specialized software for estimating model parameters efficiently, quantifying uncertainty using modern Bayesian techniques, and incorporating this uncertainty into the decision analysis (Parma 2002b, Punt and Hilborn 2001).

To evaluate policies that take into account inter-specific interactions, similar capabilities will be required from ecosystem, species-interaction, and food-web models. The challenges for scientists here are more difficult, as the uncertainty present in single-species models is compounded by uncertainty about the parameters and relationships that govern interactions among species, habitat, and the physical environment.

The simplest approach to incorporate the effects of trophic interactions into the evaluation of policies is to link the dynamics of just a few key species. An example of this approach is the model of the hake-seal system developed by Punt and Butterworth (1995) to evaluate the impact of culling the predator fur seals (*Arctocephalus pusillus pusillus*) based on the abundance and catches of the Cape hakes (*Merluccius capensis* and *M. paradoxus*). Even in such relatively simple models, the uncertainty is large. Furthermore, the decision about how to bound the system and determine which components to include is not trivial, as model predictions may be misleading due to oversimplification. For example, simple analyses of predator-prey relationships may overestimate the impact of fishing the prey species on the sustainable yield from the predator species by ignoring the complex compensatory dynamics of the entire food web. Indeed, models of the whole ecosystem tend to predict much less severe bottom-up fishing impacts than do predator-prey models because the availability of alternative prey species for piscivores buffers the reduction of some of their prey by fishing (Walters et al. 2005).

At the opposite extreme along a gradient of increasing complexity are models that represent whole ecosystems as they vary in space and time, including the environmental processes that force changes in productivity. An example of this type of approach is the model being developed to represent the California Current for the western United States (Levin 2005). The model represents marine ecosystem dynamics through spatially explicit submodels that simulate hydrology, biogeochemistry, food-web dynamics, fishing, and management. The first model of this kind was created in Australia and has already been used to help define management scenarios (Fulton et al. 2004). A socioeconomic submodel is also currently being crafted (Smith 2005) that will allow tradeoff determinations not based solely on biological concerns.

However, such models are extremely complex. The one being developed for the California Current has 1,600 parameters for each depth/spatial stratum of the model (including parameters such as growth rates, fecundity, and mortality rates). There are 64 spatial regions in the model and for each region there are 10 depth strata. Obviously, the creation of such a model, even without the addition of socioeconomic data, is an enormously data-intensive process and requires extensive knowledge of the ecosystem and its components.

Models of intermediate complexity include the increasingly popular Ecopath-with-Ecosim (EwE) models (Walters et al. 1997). These models are a dynamic extension of the ECOPATH models (Polovina 1984, Christensen and Pauly 1992), which provide a static description of the interactions between a series of functional groups. In the simplest versions of Ecosim, rates of biomass change are predicted as efficiencies multiplied by food intakes minus losses due to predation, harvest, and unaccounted mortality agents. In most applications, at least a few species are also simulated with much more elaborate "multi-stanza" accounting for size-age structure over time. EwE software is a widely used tool for the quantitative analysis of food webs and ecosystem dynamics (e.g., Pauly et al. 2000), and a number of models have been constructed to represent major ecosystems (e.g., Christensen et al. 2002; Cox et al. 2002a, 2002b; Martell et al. 2002; NRC 2003; Olson and Watters 2003; Shannon et al. 2004). The capabilities of the models and software are being constantly expanded and the impact of the various assumptions made to model trophic interactions are being scrutinized (Walters and Martell 2004, Plagányi and Butterworth 2004). Critical recent developments include the capability to input historical trends in fishing mortality or effort, productivity indices (e.g., upwelling), recruitment indices, and biomass of other, nonmodeled species to drive the dynamics of different model components. Also, predicted trends can be fitted to observed trends in relative or absolute abundances, direct estimates of total mortality rate, and historical catches (Walters and Martell 2004). Experience with fitting the model to time-series data using formal statistical methods is being gained in a few study cases, but this method is in its infancy compared to single-species approaches.

Despite growing experience with the use of multi-species models to reconstruct historical changes in ecosystems, the potential for these models to predict ecosystem responses to complex harvest management policies cannot yet be assessed (Walters et al. 2005). Still, multi-species models have evolved to a point where they can begin to provide useful tools for policy screening. For example, the Ecosim model of the central north Pacific developed by Cox et al. (2002a, 2002b) has been used to evaluate the effects of alternative strategies for harvesting tuna on the pelagic food web and the rebuilding of billfish and shark populations (Hinke et al. 2004, Kitchell et al. 2004). Management options examined include removing shallow hooks from longlines to reduce bycatch of marlins, banning of shark finning, and overall reduction of longline and purse-seine fishing effort. Simulations over 30 years indicated that elimination of shallow gear and shark finning was more effective at recovering marlins and sharks than a reduction of longline effort (Hinke et al. 2004). However, the resulting increase in predation had a negative impact on the simulated abundance and catches of yellowfin tuna, bringing about important economic tradeoffs between gains due to increased recreational catches of billfishes and major losses in tuna fishery revenues (Kitchell et al. 2004). Although several sources of uncertainty were identified, the sensitivity of the specific results to them was not examined.

A systematic evaluation of alternative model scenarios consistent with historical data would be needed if such model-based projections were to be used to inform management choices. Usually, the available data do not allow discrimination between several competing hypotheses about the structural relationships in the ecosystem. As a result, models that fit the historic data equally well may make widely different predictions about future policy outcomes. The importance of evaluating the sensitivity of predictions to model structure and parameter uncertainties is well illustrated by Koen-Alonso and Yodzis (2005) using a relatively simple interactive system involving four key species in the ecosystem of the Argentine Patagonian shelf. Several models of the predator functional responses of hake (*Merluccius hubbsi*) and sea lions (*Otaria flavescens*) preying on squid (*Illex argentinus*) and anchovy (*Engraulis anchoita*), and sea lions preying on hake, provide adequate fits to time series of abundances and catches, but make very different predictions under some exploitation scenarios. Clearly, in this example, a lack of time-series data on prey mortality rates and predator diets contributes to substantial uncertainty.

In general, food-web models may provide useful tools to simulate possible ecosystem responses as long as users are conscious of their limited predictive capabilities and therefore place due emphasis on evaluating the robustness of candidate policies relative to alternative model assumptions (Christensen and Walters 2004, Essington 2004).

Harvesting policies need to be conceived as adaptive experiments, with a requirement to implement monitoring programs to evaluate system responses and to detect unexpected consequences should they happen (Walters et al. 2005). It is only through adaptive management that our understanding of ecosystem dynamics and our ability to design robust harvesting strategies will improve. Building ecosystem models to design harvesting policies requires the cooperation of many specialists and the integration of information from many sources. This may best be achieved by a series of workshops that bring together people with different expertise. Here again, just as in the case of single-stock policy evaluation, input from stakeholders is essential for identifying important tradeoffs when evaluating feasible candidate policies.

## PROJECTING RECOVERY STRATEGIES AND THE EFFECTS OF SHIFTING BASELINES

When the Magnuson-Stevens Fishery Conservation and Management Act (MSFCMA) underwent Congressional reauthorization in 1996, greater emphasis was placed on the ecological aspects of fisheries management, including the protection of essential fish habitat, the reduction of bycatch, and the rebuilding of overfished fisheries. The Sustainable Fisheries Act (the amendment to the MSFCMA) requires not only that fisheries be managed according to the best available scientific data but also that overfished stocks be rebuilt. However, it is not clear whether the rebuilding requirement for each stock should be based on its most recent population dynamics or to some earlier dynamics. Should it be restoration to a pristine level akin to pre-Columbian contact? Pre-human contact? Or some level immediately prior to recent overfishing?

It is not known if New England cod stocks can recover to the high levels inferred from historical logbook records by Rosenberg et al. (2005), but substantially reducing fishing pressure on cod to enable recovery is a reasonable policy. Such recovery actions may be achieved by many different regulatory controls, but the critical feature of a good rebuilding strategy is its capacity to adjust in response to new information about ecosystem and food-web status by imposing more or less stringent regulations, depending on system responses.

But to be able to test the feedback capacity of any candidate rebuilding plan, it is important that a wide range of scenarios be considered in simulation trials. A major simulation study of this type has just been completed under the umbrella of the Commission for the Conservation of Southern Bluefin Tuna (CCSBT). Different decision rules proposed by teams of scientists from several countries have been tested using simulation models chosen over a series of workshops. The models span a wide range of uncertainty and imply a variety of plausible rebuilding levels (CCSBT 2005).

Simulation trials serve to quantify tradeoffs between the short-term pain inflicted by imposing immediate major quota cuts and the long-term benefits derived from possible stock rebuilding. Also, the range of likely impacts of stock rebuilding on other ecosystem components may need to be examined. For example, rebuilding for top predators may have negative consequences for their prey (e.g., the cod-lobster example discussed in the previous chapter). These are just examples of the kinds of tradeoffs that need to be explicitly evaluated to inform management choices. More conservative or less conservative decision rules can be tested, each achieving a different rebuilding target, depending on the scenario.

However, care must be taken not to limit the possibilities. As mentioned before, current models are based or "conditioned" on historical information. But what happens when our historical data has already been subject to shifting baselines? Ultimately, recovery plans should not be determined based on upper limits of population abundances estimated within the period defined arbitrarily by the date at which baseline data collection began. Scientists will need to take a long-term view when identifying the range of plausible scenarios and make use of all sources of information to protect against the shifting baselines phenomenon. In many cases, this may involve looking at unconventional sources of information that predate the establishment of regular fisheries monitoring programs and scientific databases. The "back-to-the future" approach developed by Pitcher (2001) provides a structured method for reconstructing models of past ecosystems using information about the presence and abundance of species from historical documents, archaeology, and local and traditional environmental knowledge.

## STRATEGIES FOR INFORMED AND INCLUSIVE DECISION MAKING

The previous chapter discussed mostly choices between species in managing fisheries. However, the ecosystem impacts of fishing can, and do, affect others beyond just fishermen. For centuries, humans have viewed the natural environment as a source of material resources and services as well as a source of spiritual, cultural, and aesthetic experiences. Marine ecosystems generate a diverse set of goods and services: these must be evaluated based upon consumptive uses, nonconsumptive uses, and public-good "existence" values (see below). Consumptive uses are, of course, those that rely on the removal or harvest of ocean resources, such as fishing. These uses and their value are easily quantifiable, based almost purely on market values. The value of the other two goods and services are much more difficult to measure, but that does not mean they are any less important. The most common nonconsumptive values are those related to tourism, research, and education, where their relative value depends in part on the presence of a healthy ecosystem, as well as on the cultural and economic integrity of coastal communities. Existence values are even more difficult to quantify, but are proportionally more important. Ocean and coastal ecosystems provide many services to Earth, such as climate and atmospheric regulation. These services are experienced equally by all, but few truly realize this value. This is in contrast to fisheries and other consumptive uses where a select few monetarily benefit at a much higher rate.

### Making Value Judgments

There are two main roles for public policy regarding public natural resources. One is to set regulations that define use rights for those resources. The second role is to decide macroscale allocations of public resources when different user groups come into conflict. This role has become more important recently in disputes between different consumptive users, between different nonconsumptive users, and between consumptive and nonconsumptive users.

Public policy pertaining to fisheries resources in the United States is conducted within a setting circumscribed by law. Fisheries law and other applicable environmental laws reflect ongoing and changing assertions of interest between various claimants of coastal ecosystem services. The fisheries policy process is contentious precisely because policies are not generally between right and wrong decisions, but between decisions that create both winners and losers. It is convenient to use the metaphor of a pie. Some policies (e.g., allocation across gear types) determine which group is going to get which slice of a fixed pie, namely the economic and biological yield from a system. These kinds of decisions dominate fisheries policy and they are largely zero-sum in that any policy that benefits one group will impose costs on another group. Selecting the best policy in allocation decisions primarily must reflect value judgments involving tradeoffs among different user groups. These are not science questions. However, as discussed in the previous sections, science and scientists can play useful roles by providing predictions of the likely impacts on various user groups of alternative policies, thereby making it easier for policy makers to take into account the implications of their allocation decisions.

Alternatively, some policy decisions may involve increasing or decreasing the size of the pie. These kinds of policies are particularly difficult to implement because their associated policy options have the possibility of generating win-win (or lose-lose) scenarios. For example, policies that reduce the overall productivity of an ecosystem so that sustainable harvest of all species is reduced are options that many would view as unacceptable. On the other hand, policies that enhance all dimensions of a system's productivity make it at least physically feasible to make all user groups better off without harming any particular group. These appear to be easy public policy cases because the best decisions seem straightforward. Unfortunately, win-win policy decisions are rare. Instead, most policy decisions involve gains by one group at the expense of another.

What role should science play in this decision arena where various policies affect not only the system's overall health and productivity but also who gets what share of the services provided? This is a contentious issue among ecologists and biologists and there is considerable disagreement, particularly over whether scientists ought to be recommending policies or just informing the policy process about the likely outcomes. Is it important for scientists to avoid promoting, recommending, or selecting policies? Recommendations about the "best" policy inevitably reflect value judgments about who ought to receive various slices of the public pie, whether these judgments are explicit or implicit. Judgments about which user group ought to win and which ought to lose are not scientific questions, but rather questions answerable only by adding layers of value judgments.

## The Role of Social Science and Ethics in Stewardship

Since 2002, the National Oceanic and Atmospheric Administration (NOAA) has convened a series of workshops to develop a social science research strategy to understand the human dimensions of marine protected areas (MPAs). Although the focus of the workshops was specifically on the creation of MPAs, some of the facets of planning, management, and evaluation are relevant to the development of a broad strategy for addressing ecosystem effects of fishing.

As a result of the 2002 workshop held in Monterey, California, the MPA Center identified six priority themes in social science: governance institutions and processes; use patterns; economics of MPAs; communities; cultural heritage and resources; and a category called attitudes, perceptions, and beliefs (Walhe et al. 2003). In other words, what are the variations in attitudes, perceptions, and beliefs of different actors (user groups, stakeholders, and decision makers) toward marine resources as well as toward other users? The attitudes, perceptions, and beliefs (APB) theme covers the underlying motivations that may influence human preferences, choices, and actions.

In subsequent workshops, the APB theme was elaborated more fully and generated suggestions for work germane to the topic of ecosystem effects of overfishing (National Marine Protected Areas Center [MPA Center] 2003a, 2003b, 2004). Participants suggested that it would be valuable to evaluate local knowledge about resource quality, use, access, and protection (MPA Center 2004). This inquiry should encompass perceptions of the relative condition of marine resources compared to available scientific information, and it should document elder fishermen's knowledge about changes in fishing resources over time (MPA Center 2003a). Participants suggested examining how to integrate local, traditional knowledge with scientific knowledge and vice versa. Saenz-Arroyo et al. (2005) show that these sources of knowledge are important in assessing the shifting baseline phenomenon.

Additionally, the NOAA workshops called for a determination of the differences in APBs between managers and the affected communities and stakeholder groups (MPA Center 2003a). The workshops identified the need for analysis of public participation methods in the process and how the methods themselves may affect perceptions of both the process and the outcomes (MPA Center 2003b). Participants also highlighted the need to assess perceptions of ownership of common resources between users (e.g., commercial and recreational fishermen) and nonusers (at least nonextractive users) (MPA Center 2003a, 2003b).

Two major challenges face the management community: developing social science tools, and accepting social science data as an important component of decision making (Walhe et al. 2003). While fisheries science is essential for management, by itself it is not enough. What terrestrial conservation biologists find to be true—that they manage not so much the organisms, but human behavior—holds true for the marine environment as well. To produce truly comprehensive management plans, fisheries managers must further incorporate social and economic sciences, as well as the natural sciences, in their deliberations. And part of the social science aspect involves explicit consideration of the values that underlie the decision-making process—consideration of environmental ethics.

## Defining the Role of Ethics in Management

Bringing ethics into the discourse on fisheries policy will produce a paradigm shift in the way that fisheries policy is both designed and justified. In other areas of applied ethics there have been similar shifts, the best example of which is the inclusion of ethics training in the medical school curriculum. Hargrove (1995, p. 17) noted:

> Environmental ethicists have not succeeded in developing the kind of relationship, for example, which medical ethicists have with doctors, lawyers, and policy makers. . . . Medical ethicists generally are asked to participate in the resolution of tough decisions which members of the medical community do not want to make themselves. . . . Environmental professionals have little interest in having philosophers make tough decisions for them.

Crises in natural resources management are directing scientists into the realms of ethicists and resource managers, where they now often seek ethical advice in ways similar to those experienced by the medical profession. Indeed the report of the Pew Oceans Commission (2003) called for the creation of an "ocean ethic" to guide U.S. policymaking.

Despite decades of development, the body of environmental ethics literature examining our relationship with the marine environment remains slim. Pioneers of U.S. environmental philosophy such as Pinchot, Muir, and Leopold expressed some regard for marine environments in their resource management writings, but their focus, conscious or unconscious, was on terrestrial environments. Indeed Leopold (1949, p. 251) characterized the sum of his views as "the land ethic" and wrote that "we can be ethical only in relation to what we can see, feel, understand, love, or otherwise have faith in." Where does that leave policymaking with respect to the sea? Can or should we extend the "land ethic" to the sea (Safina 2003) and to the animals who live there (Box 4.1)? How do we go about formulating and defending ethical marine policy?

Environmental ethicists who have examined the moral foundations for preservation of biodiversity suggest a shift of emphasis away from saving nature on a species-by-species basis (Callicott 1995a, 1995b; Norton 1986). This approach might prove to be even more useful for the marine environment than it is for the land. First, marine biologists may have greater difficulty than terrestrial biologists in determining which marine species to protect, particularly if they are bycatch whose population dynamics are not closely tracked. Second, by protecting habitat

> **BOX 4.1**
> **Regard for Marine Mammals:**
> **Extending the "Land Ethic" to the Sea**
>
> One of the most dramatic shifts in extending moral regard to nonhuman species is our changing view of whales (Kellert 2003). Formerly hunted by the hundreds of thousands to produce commodities such as lamp oil and margarine, whales and other cetaceans have attained an extraordinary level of ethical concern within the time span of a single human generation. Granting some important exceptions, whale killing has now been supplanted by whale watching as a source of income as well as an aesthetic and naturalistic experience (Russow 1981). In the United States, the Marine Mammal Protection Act of 1972 created a moratorium on the taking of marine mammals not only in U.S. waters but also outside U.S. jurisdiction— prohibiting such activity by any U.S. citizen on the high seas. This law provides a stringent code of ethics with strong penalties for violators and with very narrow exceptions. Additionally, consumer demand for dolphin-safe tuna led the U.S. government to impose bans on importation of tuna caught using fishing techniques that resulted in high dolphin mortality.
>
> But what about other marine organisms? Compared with their terrestrial counterparts, marine species are much less protected. <u>Few truly marine animals and plants appear on the U.S. endangered and threatened species list</u>, and there seems to be even less support at the international level for adding marine species to international lists such as those promulgated by the Convention on International Trade in Endangered Species (CITES) of Wild Fauna and Flora (CITES 2005).

integrity, a whole range of organisms is protected as well as the integrity of their interconnectedness.

There are two aspects of utilitarian ethics that could be used to modify behavior in ways that could result in more favorable outcomes for the sea: One is how inclusively the "self" in "self-interest" is defined and the other is the time frame considered (Tam 1992). The definition of self can be thought of as an issue of scale, a concept familiar to conservation biologists who study ecological scales that range from genes to the biosphere. A multi-scalar analysis reveals that human impacts occur at many different spatial scales (Norton 2003). If the ethical scale is extended to include a moral regard for the biotic community, then it is possible to consider not only impacts of human activities on other humans but also on marine domains of all sizes. As a consequence, an opportunity opens for a wider-ranging discussion of the multiple ethical values which are underpinning the duties of stewardship that humans owe the marine environment.

Changing one's scale in time, as well as space, also can modify the perception of what best advances self-interest. Leopold (1949) counseled his readers to "think like a mountain," and thereby extend the time reference from experiential to that of ecological (or even geological) scale. Extending the time horizon requires users of the marine environment to balance perceptions of what is currently in our self-interest against our potential foreclosing of environmental options that might be available to future generations (Knecht 1992). This concept, one of intergenerational equity, is grounded in part on an ethical stance requiring persons to act in such a way that they would want their actions to be universalized. As Rawls (1971) characterized it, how would our actions change if we did not know which generation we are in?

Including consideration of future generations in current policy decisions does not mean we try to forecast their preferences. But we can scientifically assess the fragility of marine ecosystems, and we do have the technology to alter them irreversibly. According to Norton (1991, p. 219), "The lesson of ecology is that one cannot care for the future of the human race without caring for the future of its context. . . . Context gives meaning to all experience; consequently, it is a shared context that allows shared meaning—what we call culture—to survive across generations." In Norton's sense, then, the sea ethic we create must link past, present, and future generations in a culture that recognizes and respects limits on our actions in the sea.

## Public Representation in Fisheries Management Decisions

The Magnuson-Stevens Fishery Conservation and Management Act created eight Fishery Management Councils as administrators for the living marine resources within the U.S. exclusive economic zone. One of the principal goals of the Act was to link the fishing community more directly to the management process. In fact, the Councils are a principal avenue for public involvement in deciding fishing limits and regulations. To date, the majority of public appointees are fishermen and fishing industry representatives (both commercial and recreational); there have been few appointments from outside the fishing community. Granted, this "unequal" representation might not have been such an issue when the resources were thought to be boundless and there was no recognition of the role fisheries could play in shaping the functioning of whole ecosystems. However, armed with this new knowledge, methods for bringing other interests into the decision-making process may be warranted.

Recently, NMFS has been experimenting in some regions with a system designed to improve access and involvement of multiple user groups and stakeholders as stock assessments are developed. For example, in the Gulf of Mexico and South Atlantic regions, this process is called SEDAR (Southeast Data and Assessment Review) and in the Northeast, SAW-SARC (Stock Assessment Workshop-Stock Assessment Review Committee). The new process employed by these groups includes a series of workshops where NOAA scientists, fishermen, nongovernmental organizations, industry consultants, academics, state biologists, economists, and others meet to discuss the data sets needed to develop an assessment. This is followed by a second workshop where a similar diverse group (with some overlap of members) builds on the output from the first workshop and produces the assessment. Each assessment is reviewed by an independent group of experts before it is passed to the Council's Science and Statistical Committee. While this process has increased the number of meetings—and therefore the time—it takes to produce final products, the assessments created through this process are generally better received by the Councils and their advisory panels, and skepticism about the assessments is reduced. The apparent success of these new organizations supports the implementation of a similar approach to incorporate stakeholder input when discussing ecosystem impacts of fishing and when examining alternative model-based scenarios for management actions.

Creating a formal approach for evaluating alternative management strategies using food-web models, as discussed in a previous section, will not be a productive exercise if those affected by alternative actions are not involved in the decision-making process. Obviously, choosing among tradeoffs and settling conflicts between two competing fish species is not the greatest issue. If we are to include in our scenarios ecosystem recovery options to develop other uses, or to decide tradeoffs between consumptive and nonconsumptive uses, the proponents of these uses cannot be excluded from the discussion. This is particularly true if cascading effects or trophic interactions caused by fishing will prevent or compromise these other uses or services.

## MAJOR FINDINGS AND CONCLUSIONS FOR CHAPTER 4

**More extensive use of food-web and other ecosystem simulation analyses are needed to explore possible consequences of different candidate harvesting strategies under alternative scenarios representing the state and dynamics of marine ecosystems.** Fisheries management advice has tended to follow prescriptive policies defined in terms of generic biological reference points for individual populations. However, within an ecosystem context, tradeoffs between conflicting management objectives need to be made explicitly by evaluating policy consequences in terms of different measures of performance that reflect policy impacts on various ecosystem components and uses, including consumptive and nonconsumptive uses.

**Owing to their inherent complexity and associated uncertainties, ecosystem models are unlikely to provide numerical tactical advice on fisheries regulations.** The main use of ecosystem models in the near future will be to build alternative scenarios to test strategic policy choices. The challenge for scientists and managers is to identify and assign probabilities to a range of scenarios that captures existing uncertainties about the food-web dynamics and the responses of fished food webs to various fishing strategies.

**Food-web and other ecosystem models currently exist that provide useful tools for policy screening.** Not all interactions need to be known to begin creating and applying existing models of key species interactions and food-web components and to aggregate and manage the underlying data.

**Scenario analyses and the corresponding management actions are best applied in an iterative and adaptive process.** As management is applied and knowledge about marine ecosystems increases, some scenarios will be seen to be less plausible while other new scenarios may emerge. In addition, new information about how humans respond to regulations may favor some policies more than others. The models themselves will improve as more is learned and greater levels of complexity are added, requiring an adaptive approach to management.

**A diverse cross-section of constituents may be needed to weigh the varied ecosystem values and uses involved in model-based scenario analysis.** An important public policy issue is how to assure that nonconsumptive and public-good values receive proper consideration when making tradeoffs among ocean services. Commercial and recreational uses are represented through the Fishery Management Councils, but the current composition of the constituents "at the table" does not represent concerns of potential ecosystem services beyond fisheries extraction.

# 5

# Science to Enable Future Management

Many ecological and fisheries publications have raised the issues of ecosystem effects of fishing on populations, food webs, and communities. While the evidence presented in Chapter 2 is compelling, it is by no means conclusive about all possible effects. There is still much that we do not know.

To comprehensively understand the ecosystem effects of fishing, we must know the current state of the ecosystem, the state of the system at earlier stages (preferably pre-exploitation), and what factors contribute to stability or change. Additional research is needed to underpin the incorporation of ecosystem effects of fishing into fishery management and to help decide the allocation tradeoffs discussed in the previous chapters. If we are to make tradeoffs between uses and between species, we must try to anticipate the possible outcomes of candidate management policies. Given our present understanding, not enough is known to "steer" exploited marine ecosystems. The ability to move a whole ecosystem toward a desired dynamic state by altering fishing targets, catches, and effort is limited because many variables are unknown and many ecosystem drivers are not under our control. These variables can be unique to each system; managers must appreciate the level of ignorance about the systems, the limits of our forecasting capability, and the consequent uncertainty generated by capricious, extreme events.

Data perform a central and dual role as indicators of ecosystem performance, while also providing input variables to complex models, such as those representing ecosystem dynamics and management scenarios used to assess tradeoffs in decision making. New data can challenge existing theory and facilitate wider application of model-based scenario analyses; but changes in what we collect and how we manage data will be necessary. Finite resources require us to reassess available data and to revisit existing monitoring programs and data resources. However, we must prioritize data needs both for near- and long-term efforts with complex, multiple objectives.

Promising results have come from analyses and models at levels of synthesis above individual populations and individual food-web components. Some of the data, models, and knowledge are now sufficiently developed to be applied to evaluating ecosystem approaches to management. However, the derived management actions inevitably will be experiments in themselves—adding to a growing body of knowledge and creating an environment of adaptive management.

Clearly, moving forward requires science, management, and policy interacting constructively in synergy. We need to think outside of the box: incorporate new ideas, new analyses, new models, and new data, and perhaps most importantly, establish the social and institutional climates that will catalyze creative, long-term, comparative, and synthetic science of food webs and communities applied to exploited ecosystems. Data needs in support of ecosystem-based management will likely be more than the simple sum of currently available single-species information. Diet data and strengths of linkages between species and life-history stages will be as important as population abundance data. A rich array of social science, economic science, and policy considerations is essential because many more tradeoffs among ecosystem components and stakeholders are likely to be apparent. Science will be challenged to provide policy-relevant options in this new context; managers will be challenged to broaden their concerns and experiment openly; and policy makers will be challenged to act unselfishly in behalf of the broader community of people who value and depend on ocean ecosystems.

This chapter presents the need for research on food-web interactions, spatially explicit data, complete historical time series, and scientifically useful definitions of ecosystem boundaries. There is also discussion of future needs in valuation of nonmarket services, fishing behavior, and integrated bioeconomic modeling.

## IMPROVING ECOSYSTEM MODELS AND SCENARIO ANALYSIS

Choosing goals and standards that would be appropriate for food-web and community management is a worthy but formidable challenge. As discussed in the previous chapter, such approaches will be important in any comprehensive fisheries management planning and will likely include consideration of a wider variety of ecosystem services than the food and economics of fish yields. However, if model-based scenario analysis is to be used more extensively in fisheries management applications, many variables must be better defined and understood to reduce the inherent uncertainty.

Models in marine science and fisheries range from whole-system ecosystem models to single-species population models; many have been around for decades, at least in some form. A variety of modeling approaches are now available to address the dynamics of marine food webs from multi-species fisheries models (May et al. 1979, Hollowed et al. 2000), to Ecopath/Ecosim/Ecospace (Polovina 1984; Walters et al. 1999; Pauly et al. 2000; Christensen and Pauly 1992, 2004). While some of these have received more attention and application than others, there will certainly be use for many different kinds of models, including those based on the underlying knowledge of the systems, the extent of resource exploitation, the assumptions made within the model architecture, and the goals for management.

## Food-Web Interactions

Currently, the most popular multi-species models focus on community ecology and food-web interactions (Crowder et al. 1996, Mangel and Levin 2005). This approach avoids the problems of managing single-species populations as though they can be isolated from key species with which they interact, particularly their prey and predators. But the emphasis on community ecology and food-web interactions also avoids building in the overwhelming complexity of marine ecosystems, including interactions and linkages that may or may not be important to the dynamics of species under management.

Food-web modeling approaches can focus on all food-web connections, or on those that transfer the largest proportion of energy and materials, or on those that exhibit strong interactions (*sensu* Paine 1980) in which the "unit" is made up of individuals. Although interaction strength in the field initially was quantified only by manipulative experiments (Paine 1980), insights from a large number of experiments and "natural experiments" driven by climatic variation (Francis et al. 1998) or heavily exploited fisheries have revealed strong interactions in some marine ecosystems (Frank et al. 2005). It is currently unknown whether the pelagic trophic cascades of Estes et al. (1998) and Frank et al. (2005) could be modeled, and hence predicted. But with increased understanding of per capita effects or population effects, it may be possible to account for the dynamical changes at a variety of trophic levels, and thus legitimize the concept of ecosystem-based management.

Estimates of per capita interaction strengths have been made often over the years. They were mathematically formalized by Lotka and Volterra in the 1920s (Lotka 1925, Volterra 1926), led directly to Gause's (1934) experiments, and then led to a discussion of application in MacArthur (1972), Paine (1992), and Laska and Wootton (1998). In a completely described trophic interaction, the reciprocal effects of predator on prey and prey on predator are quantified. Such information would permit questions on how changes in predator and prey abundance influence their respective fecundities and mortalities. Numerical and functional responses can be considered as can the consequences on predator performance or predator switching (Murdoch 1969). Further, it is plausible that the ecologically important consequences of ontogenetic changes in predator diet (Hardy 1924, Hutchinson 1959) can enter quantitative ecosystem models.

Can the necessary estimates be developed in a manner useful to the management of fisheries? Most signs are positive. Wootton and Emmerson (2005) review the evidence that studies on per capita effects typically identify a few strong and many weak interactions. The question of whether such information justifies the simplification of food-web studies to a limited number of "important" species is not trivial. More significantly, two recent studies suggest that per capita estimates can be obtained from combined field observations and measurements; experimental manipulation is not necessary. Wootton (1997) develops this approach for a rocky intertidal assemblage in which three bird species consume at least twenty-three taxonomically varied prey. Bascompte et al. (2005) examine a Caribbean food web of 249 species dominated by fishes; interaction strength is calculated as prey biomass consumed per predator per day. However, theses two studies were only possible because the web architecture and species diets were known. Both of the studies involved a significant benthic component and were accessible to direct observation at multiple trophic levels. As such, many of the limitations imposed by equally complex but entirely pelagic or less well studied systems were avoided.

But it is important to note that these food-web interactions can change drastically over time. In fact, the Bascompte et al. study (2005) analyzes stomach contents of fishes from the 1950s and 1960s. Yet species commonly used as food by large predators such as sharks and groupers have virtually disappeared from Caribbean reefs, so the topology of links as well as interaction strengths are possibly very different today. Thus, even if these large predators recover, their diets will most likely be different. These concerns underscore the need for improved data on evolving diets due to changes in connected species.

Many models simply ignore, of necessity, the generally unknown ecological complexities characteristic of all lower trophic levels. The consequences of aggregating many seemingly comparable taxa into "trophospecies" are unknown, but it artificially simplifies the specific relationships in the food web and potentially disguises important factors. Yodzis (1982) and Martinez (1991) explore the consequences of aggregation in real food webs. The dynamics of lower trophic levels on which the fished species depend is often poorly known. For instance, Ward and Myers (2004) report increases in squid and fishes (pomfrets and midwater sting rays) in mid-Pacific areas where apex predator presence has been reduced. Does this hint at a possible cascade? Is there a measurable ecological effect of an increase in these secondary prey species? There can be no answer in the absence of data on these species, such as population abundances or their individual natural histories.

Data on age specific diet continue to be a priority because this information is essential to identifying significant trophic linkages and per capita interaction strengths. Equally vital are natural (and fished) mortality rates obtained from standard surveys and tagging studies. In addition, the per capita interaction strength often needs to be calculated in biomass units to reflect the changes in size structure imposed by fisheries or simply interspecific differences. So another concern is that although trend data are available for landings of many fished components, relating landings to biomass in the ocean remains a problem.

## Tying Data to Spatial Scales

Interactions in the marine environment occur in a spatial-temporal matrix, and strengths of food-web interactions depend on the species being in the same space at the same time. Fisheries scientists usually deal with large spatial scales, whereas the relevant data are obtained from finer, spatially explicit regions. While it may have been sufficient to substitute variability observed over time in matters concerning often relatively local single-species management approaches, it will be necessary to characterize variability both in time and space to reduce the uncertainty at the ecosystem level. Space has been called the final frontier in ecological theory (Kareiva 1994), and spatial analyses may be one of the greatest obstacles faced by fishery managers (Caddy and Garcia 1986, Garcia and Hayashi 2000).

Developments in measurement and methods of analysis make explicit consideration of spatial variability possible. The general availability of global positioning systems permits much greater accuracy in determining where samples are collected than previously possible. Geographic Information Systems (GIS) provide a means for storage and interpretation of spatially explicit data (Keleher and Rahel 1996, Johnson and Gage 1997, Wiley et al. 1997, Clark et al. 2001). Remote sensing allows simultaneous measurement of environmental variables over large scales (Cole and McGlade 1998, Polovina et al. 1999). The recognition that population dynamics are influenced by large-scale oceanic patterns and decadal climate patterns such as the El Niño-Southern Oscillation and the Pacific Decadal Oscillation has aided in explaining variation and trends that may appear random at a local level. At smaller scales, animal positions and movements can be related to fixed and dynamic oceanographic features (Magnuson et al. 1981), but improvements in current tools for marine geospatial analysis are needed to achieve these goals.

Ecological data show that context is extremely important. The ability to tie landings and fisheries-independent data to finer spatial scales will allow researchers to examine local population trends and localized depletions. Furthermore, better understanding of the spatial distribution of stocks, life histories, and species interactions may allow for a new generation of management techniques based on this new knowledge, including that gained from the creation of transient protected areas.

## Defining Ecosystem Boundaries

Complex biological systems cannot be reduced to a single diagnostic scale (Levin 1992), thus multiple spatial scales are applicable in research and management endeavors. Ecosystem-based management acknowledges the significance of a multi-species perspective and implies that the spatial boundaries defining the system of choice can be identified. The concept of an "ecosystem"—roughly defined as a biotic assemblage and its environment—was originated by the English botanist A.G. Tansley and given its modern face by R.L. Lindeman (Golley 1993). This conceptual evolution derived from the study of partially closed, physically bounded environments such as lakes, ponds, grasslands, and forests. But application of this definition remains problematic in the marine environment due to physically large systems that lack distinctive boundaries and are "open" in the sense that nutrients and species can be exchanged over great distances.

The growing recognition of important exchanges between distinctive ecosystems further blurs the concept. Landscape ecology (Turner et al. 2003) recognizes the importance of interactions among "patches," which may be a more reasonable construct. For example, migratory species move through and between identifiable parts of the ocean. Nutrient-laden terrestrial run-off impacts marine coastal ecosystems (Rabalais et al. 2002), and marine nutrients transported by birds to Aleutian Islands power the productivity of those systems (Croll et al. 2005). The increasingly recognized reality of such "subsidies" (Polis and Hurd 1996), transported across the bentho-pelagic boundary, represents yet another challenge to management of marine resources.

The importance of the term "ecosystem," as in ecosystem-based management, is that it acknowledges numerous interacting species that must be managed simultaneously. Defining precise marine ecosystem boundaries is a biologically unrealistic, unattainable goal. Instead, ecosystems may be better defined using knowledge of food-web structure and embedded interactions, with the boundaries determined by management goals, financial constraints, and other realities imposed by political considerations.

## Applying Models to Management

Testing new modeling approaches in food webs and communities where fishing occurs is an important endeavor. The challenge in ecology and fisheries is to forecast ecosystem behaviors in response to management manipulations when we are unsure of the baseline and have little idea of what the future will look like. However, the alternative to no action and walking away from the issues owing to limited data or inadequate models is not acceptable. Iterative development—that includes trial, monitoring, and eventually feedback—will most certainly be needed.

Building models relevant for fisheries management requires the cooperation of many specialists and the integration of information from many sources. Most likely, this is best done over series of workshops that bring together people with different expertise. Workshop participants with interests beyond the fishing industry will help to bring broader management objectives to the design of ecosystem management approaches. Furthermore, just as is the case of single-stock policy evaluation, seeking input from a wide variety of stakeholders is essential for identifying important tradeoffs to evaluate the feasible candidate policies.

Similar working groups can be used as a template for the formation of such modeling endeavors. For example, the National Center for Ecological Analysis and Synthesis (NCEAS) is currently developing a modeling and data integration framework for ecosystem-based management and applying that framework to a case study from coastal California. NCEAS is connecting experts in the modeling of natural and human systems with policy specialists to forward the goal of developing a policy-relevant modeling approach that includes the dynamics of social, biophysical, and economic components of the ecosystem and critical feedbacks among them. The Long-Term Ecological Research system sponsored by the National Science Foundation has also conducted similar workshops, which gathered investigators from several distinct sites to model and test new ecological analytical techniques.

## ANALYZING HISTORICAL TIME-SERIES DATA

Current data gathering and analytical approaches for fisheries management will be necessary but not sufficient to support effective management of fisheries impacts on food webs and communities. Further progress is being made through studies that are reconstructing the history of exploited ecosystems, using sophisticated and creative data analysis and synthesis tools to extend data well back in time (Pauly et al. 1998a, 2005; Myers and Worm 2003, 2005b; Jennings and Blanchard 2004; Rosenberg et al. 2005). The focus of these studies has been on the history of fishery impacts, which provides only part of the information required to develop a greater mechanistic understanding of ecosystem function.

Because of their availability and length of time series, landings data are most often used for both historical, retrospective analyses and current multi-species, dynamical ecosystem models. Often the sampling extent is long; for instance, landings of five species of Alaskan salmon date back to 1880 (NRC 2003). However, numerous problems are recognized; landings may or may not reflect population numbers in the wild, can be subject to often unknown effort, and will be influenced by changing demand and catch technologies. Nonetheless, such data sets are increasingly acquired, resolved to varying degrees, and made accessible to interested parties.

Historical data—both long time-series and specific snapshots—have recently been used to examine severely reduced components of food webs and to evaluate the potential linkages of these components to consequent population changes. These data can allow glimpses into the past, possibly even pre-exploitation. Thus the roles particular species played in food-web dynamics prior to their decline can be examined. Including these time-series or snapshot data in ecosystem and food-web models may provide the best approach to synthesizing long-term data and identifying alternative future scenarios to evaluate policy choices.

Frank et al. (2005) provide a second example of research integrating disparate time-series data in their description of combining a trophic cascade involving Atlantic cod (data from landings), pelagic fishes, shrimp and snow crab (catch numbers standardized by effort), zooplankton size structure (from plankton recorders), phytoplankton (from a color index), and nutrients (nitrogen). Analysis and integration of these types of data will not only continue to inform managers about the dynamic challenges they face, and the consequences of mismanagement, but will also be invaluable for the application of ecosystem models to novel systems.

Some of the other efforts to recover information of past population sizes and distributions have generated controversy (e.g., Springer et al. 2003, Palumbi and Roman 2004). Questions always remain about their interpretation; for instance, how accurate are Jackson et al.'s (2001) historical reconstructions, or how abundant were Atlantic cod 200 years ago (Rosenberg et al. 2005)? A variety of techniques are involved, some qualitative (e.g., site photographs), others quantitative (e.g., standardized time series). None are without flaws, but the quest to reveal historical levels of exploited populations is vital to determining baselines around which to establish fisheries management and recovery goals.

## Conserving and Accessing Data

An information system is needed that increases access to historical data, incorporates data from disparate sources, and supports the policy-making process. Access to data from diverse sources will facilitate the transformation of these data into useful information that leads to model-based scenario analysis and informed decisions. Large-scale modeling of marine ecosystems requires facile integration with data from multiple sources. In addition, new sensors are being incorporated into ocean observing systems deployed across large scales that are capable of generating massive data streams. The use of these data will demand the appropriate information technology infrastructure for data storage, management, access, and analysis, often in near-real time. Data streams that incorporate fine-scale, spatially explicit data will be particularly useful to improve understanding and to enhance management. Synthetic scientific endeavors can be hindered by the difficulties inherent in discovering and accessing data across heterogeneous systems.

Furthermore, collaboration technology is needed to support interactions in science and management. Information technology infrastructure is an essential component for generating ecosystem and food-web analyses for policy scenarios. Ideally, information systems would provide the scientific community with data discovery and access capabilities as well as the documentation of the data that fosters the interpretation and integration of multiple data sets.

Development of an information technology infrastructure that would provide a highly functional platform for scientific endeavors requires meeting challenges in a number of areas: data discovery, access, evaluation, and integration; a framework for modeling that provides repositories and documentation for models and model output; frameworks for designing and executing complex workflows; and advanced analysis and visualization tools, particularly for spatially explicit data. A prior group that examined future data needs of ocean science identified the following issues that need to be addressed: (1) technical support for maintenance and upgrade of local information technology infrastructure resources; (2) model, data, and software curatorship; and (3) facilitation of advanced applications programming (OITI Working Group 2004).

## CONTRIBUTIONS FROM SOCIAL AND ECONOMIC SCIENCE

In the United States and in other countries with modern fisheries management systems, the task of managing just a single species should require as much information about socioeconomic aspects as it does information about biological mechanisms. As managers gradually adopt multi-species and ecosystem-based management methods, the social and economic information needs evident in current single-species systems will be amplified. Thus, there is a need for new research from the social and economic sciences to better inform future management processes. Three areas of particular interest include valuation studies, integration of biological and economic models, and examination of governance options for managing ecosystem goals.

### Valuing Nonmarket Ecosystem Services

One of the most important under-researched areas pertaining to marine ecosystems is the issue of ecosystem services valuation. The notion of ecosystems services is a broad and encompassing term, intended to include more familiar marketed commercial services such as the value of fish harvested, but also the value of components and characteristics of ecosystems that are not consumed or marketed (for a taxonomy, see Grafton et al. 2001). Understanding and measuring the values of actual and prospective portfolios of services is critical to making sound policy decisions about the use of the marine environment (Lange 2003). Various social science disciplines have focused attention on how humans form values, how those values change, and how values from different experiences compare and scale vis-à-vis one another. Economics has a well-developed methodology for measuring and scaling economic values that borrows from cognitive psychology and survey research theory (Freeman 1993, NRC 2004). Much of the valuation research over the past decade has been devoted to measuring nonmarket values, or values that humans place on ecosystem services that are not tied to any market transactions.

While there have been numerous applications of nonmarket value methods to the services provided by terrestrial systems, there are few studies devoted to marine systems (Barbier 1994, Swallow 1994). The kinds of questions that need addressing are: how do the values provided by on-site, nonconsumptive services (e.g., diving, tourism, education, research) compare with on-site consumptive services (e.g., commercial and recreational harvesting, oil and gas production, seabed mining)? What are the determinants of off-site nonconsumptive service values (e.g., existence values, posterity benefits)? How do humans perceive the value of intact and relatively pristine marine environments? What intrinsic characteristics of systems do humans value; what losses are perceived when systems lose those functions and characteristics? How do human values relate to the rarity and uniqueness of ecosystems and/or assemblages of species? What human values are associated with stability, resilience, diversity, and other similar characteristics of intact systems? These are but a few of the issues that have not been explored but that need further investigation in order to make informed social decisions about the possible outcomes of fisheries policies.

## Fishermen's Behavior and Fleet Dynamics

As ecosystem-based management methods are implemented, there will be a significant need to model and predict the behavior of fishermen and to integrate this information into ecosystem-based biological models. The "second generation" of ecosystem models must account for changes in fishing capacity—including the effects of economic conditions, changes in climate, changes in market and trade conditions, advances in and increased costs of technology, and changes in abundance (e.g., those changes associated with biological interactions and regulatory changes).

Minimal research exists on determinants and mechanisms of fleet behavior. There is some work on coarse-scale, long-horizon choices made by fishermen and some research on fishermen's aggregate entry/exit behavior in fisheries (Wilen 1976, Bjorndal and Conrad 1987). Results from empirical studies of entry/exit (fleet size) behavior under open access confirm that fishermen enter when current and recent profits are positive and exit when profits are negative. Even fewer studies have examined capacity creep or "capital stuffing" when access is closed because of limited entry (Pearse and Wilen 1979).

Economists have studied the determinants of and speed at which fishermen switch target species and gear use over various time scales from daily decisions to between-season decisions (Bockstael and Opaluch 1983). These studies show that the main determinants of switching decisions are profits, so that price changes play as much a role as catch rates in determining target species and gear. The studies also find sluggishness in the adjustment between species/gear combinations in the sense that targeting and gear switches take time to fully unfold. Individual micro-level behavior at fine time scales (e.g., daily or weekly) is modestly responsive to short-term catch and price changes, but much more responsive over longer periods when catch rates and price changes persist.

Understanding the spatial behavior of fishermen will also become increasingly important. In one of the first empirical studies of spatial behavior, Hilborn and Ledbetter (1979) conclude that a British Columbia purse seine salmon fleet was composed of a mobile fleet that adjusted rapidly to changes in revenues over space and a sedentary fleet that seemed to enter and participate when revenues exceeded some threshold level. Eales and Wilen (1985) show that, in repeated daily decisions, pink shrimp fishermen select patches in a manner consistent with forecasts of average revenues based on the previous day's data. The same study shows shrimp concentrations to be ephemeral and that fishermen use information-sharing mechanisms to expand their spatial search. A few studies use panel data on a very fine scale (daily or hourly) to study short-term location and fishing participation choices. Smith and Wilen (2003a) also estimate a model of daily location and participation choice by sea urchin divers, exploring differences between short (daily) and longer-term (monthly) spatial behavior. Spatial responsiveness increases in the long run as fishermen switch home ports and home regions (Smith and Wilen 2003b).

Understanding how fishermen react and behave as a result of different management actions is essential when developing ecosystem-based management methods and solutions. Investigations of behavior, changing in both time and space, are greatly needed in order to parameterize models of fishing effort determination. The social and economic models within ecosystem-based management need to become as robust as the biological models in that system. Furthermore, better understanding is needed of how behavior and choice affect the interactions of fisheries and other sectors. This knowledge could lead to better decision making on the tradeoffs between sectors and uses and could create greater acceptance of regulatory measures.

## Integrated Bioeconomic Modeling

Moving toward rational assessment of various ecosystem-based management alternatives requires understanding both the biological linkages connecting various species and the environmental medium, and also the manner in which humans have an impact on systems and the values placed on various impacts. This understanding will emerge only by developing and examining integrated biological-social-economic models, or models that explicitly link biological and human modules. Two different modeling approaches need further development. The first is what might be called impact analysis modeling. Impact analysis attempts to measure the first-order consequences on humans of various policy-induced changes in ecosystem characteristics. An example would be a simple cost/benefit computation of the effects of various restoration strategies in a multi-species system where the interconnections are reasonably well known. Impact analysis generally assumes passive response of humans to policies.

A second, more sophisticated kind of integrated bioeconomic modeling approach incorporates active behavioral assumptions about humans (Wilen et al. 2002). For example, one might model fisherman's fishing location choices with a model that incorporates the economic motivations behind those choices, and then use that model to predict how fishermen would reallocate after establishment of a marine reserve (Holland 2000, Smith and Wilen 2003a). Behavioral models allow analysts to predict the biological and economic consequences of policy measures by understanding how human behavior is altered with policy changes, and how those behavioral changes impact ecosystems. Few fully integrated bioeconomic modeling studies have analyzed policy options using behavioral modeling approaches (Carpenter et al. 1999, Carpenter and Brock 2004).

Furthermore, little integrated bioeconomic analysis has been based on the fundamental production relationships among interacting species and between species' health and habitat conditions and characteristics. We know little about how disturbance and destruction affect the fundamental production relationships among dependent organisms. For example, how does urchin harvesting affect groundfish via its affects on kelp forests? This kind of understanding is critical to more effectively manage competing consumptive service production from ecosystems. Aside from habitat connections, we know little about even simple multi-species interrelationships in systems that produce multiple and competing marketed services. For example, how does the commercial harvesting of menhaden for fish oil affect participation and valuation per recreation day in the valuable sports fishery for striped bass? How does nutrient pollution and runoff from agriculture affect higher-level trophic organisms via their link to lower level organisms in dead zones? These are all questions that involve linkages between components of marine ecosystems, which then translate into (mostly) on-site consumptive market value services. Understanding them requires integrated bioeconomic modeling and calibration in ways that capture important biophysical linkages, translated through to human impacts via economic market valuation methods (Söderqvist et al. 2003).

## Institutional Options for Ecosystem-Based Management

Little research effort has been devoted to answering what kinds of governance and management institutions are best for conducting ecosystem-wide management for marine systems (Rudd 2004). As the Pew Oceans Commission (2003) and the U.S. Commission on Ocean Policy (2004) reports suggest, governance issues will be critical for adopting ecosystem objectives and for implementing policies to carry them out. A wealth of related thought in the social science and management science literature addresses components of this question, but less research puts issues in the context of marine ecosystems specifically. In essence, marine ecosystems can be thought of as having the capability to generate an almost infinite variety of ecosystem services, depending upon the structure and characteristics of the system. But society can alter ecosystems and ultimately choose the portfolio of services provided from among a feasible mix associated with direct consumptive services and nonconsumptive services (van Kooten and Bulte 2000).

The more important dimensions of this question for fisheries are those that deal with accounting for species interactions, incorporating nonconsumptive use values, making tradeoffs among and between user groups, and generating stewardship values (Ferraro and Kiss 2003) among users of consumptive services. These, for the most part, are extensions and elaborations of issues currently confronted by managers guided by single-species approaches. It would be speculative and probably of no use to point to specific, finely defined research that would solve these institutional design problems. On the other hand, a variety of institutions currently operating in various parts of the world are tackling many of these problems, with varying degrees of success. Many of these are conventional top-down systems, but a large number of experiments with decentralized systems are also in progress, including individual transferable quotas (ITQs), individual quotas (IQs), harvester cooperatives, community cooperatives, and territorially defined cooperatives. At a minimum, a sensible first step would be to carefully summarize experience with other management structures, including those that are devolved down to varying levels of local control. Much can be learned from existing examples of rights-based systems, many of which are currently also incorporating nonmarket values (like their top-down counterparts) and addressing issues such as interspecies conflicts and quota setting, voluntary interspecies quota trading, market based bycatch schemes, and habitat damage reduction measures.

## MAJOR FINDINGS AND CONCLUSIONS FOR CHAPTER 5

**Greater knowledge of food-web interactions, including interactions at lower trophic levels, will be essential to improving ecosystem and food-web models.** Model development is based on knowing species interactions and the strength of these interactions. By necessity, many species are ignored in current models or are grouped together based on trophic level. More research is needed to determine whether these simplifications are sound, and whether current models have ignored species with strong interactions in the system. Collecting baseline data on a number of non-target and lower trophic level species may also aid in model development. Without these data it is difficult to determine the role of these species in the ecosystem when their abundances and interactions change due to fishing pressure.

**The development of new ecosystem and food-web models will be a highly interactive process that will require input from many disciplines.** Collaborations of numerous scientists, managers, and stakeholders in a workshop setting will most likely be the best approach to develop, test, and apply new models. The incorporation of social and economic scientists at the beginning of this process will ensure that these issues are not left until after the biological model is created.

**Combining biological and spatial data will allow both large-scale population trends and changes at finer scales to be monitored and understood.** Patterns of interaction, and the strength of these interactions, vary in time and space. If data are collected in both dimensions, the models created for management scenarios will better account for this variability. In addition, understanding interactions on various scales can help to define what the ecosystem might be. Although the term ecosystem is in general use, specific boundaries in the marine environment are difficult to identify. Rather, defining an "interaction space" with boundaries determined by nonecological considerations may be a more workable solution.

**Assessing historical data can continue to lead to new insights about former species abundances and interactions.** Comprehensive analyses of existing data can reveal ongoing changes in target species as well as habitat and non-target species that are thought to indicate ecosystem status. Landings data, narratives and descriptions, fisheries-independent data, phytoplankton records, satellite data, and archived specimens can all be evaluated to develop insights about prior ecosystem status. Data need to be made available at the highest possible spatial resolution. Highly aggregated fisheries data do not allow appropriate integration with environmental data.

**Currently, fisheries data are fragmented and dispersed, which is slowing the use of these data in comprehensive analyses.** If historical data are to be combined and reassessed in new ways, it is essential to collect and provide open access to these data. It will be particularly important to access a wide variety of data and information when developing and applying model-based scenario analysis. It is important not to underestimate the need for information management support for collaborative efforts to create workable models.

**Future protocols for ecosystem-based fisheries management will place new demands on social and economic analyses to determine tradeoffs and make strategic decisions.** If value-based tradeoffs are to be made when determining fishing harvest strategies, there must be an understanding of the value assigned to non-target species, ecosystem functioning, nonconsumptive uses, and large-scale processes such as climate regulation and nutrient cycling. Furthermore, combining biological and socioeconomic information into integrated models will allow for the explicit consideration of these sometimes conflicting values.

**Various management institution options are available that change the race-to-fish incentives for fishermen, and encourage stewardship in single-species systems.** Individual quotas, harvester cooperatives, community cooperatives, and territorially defined cooperatives exist in a handful of fisheries in the United States and in other countries. But research is needed to understand how these systems affect incentives in a multi-species setting, and how they might be adapted to handle more inclusive goals associated with fisheries management in the United States.

# 6

# Findings and Recommendations

Challenges in assessing and responding to the ecosystem effects of fishing include gauging the magnitude, spatial extent, and mechanisms of change in marine food webs. Identifying and understanding these potential impacts and interactions will be essential for developing future management actions. However, better scientific understanding is not enough. Stewardship of the marine environment will demand difficult societal choices and tradeoffs between uses, because resources are connected through food-web and ecosystem interactions; maximizing each individual harvest or use may be impossible as the upper limit of available productivity is reached.

Science has revealed numerous ecosystem-level effects of fishing in marine ecosystems. There is conclusive evidence that stock biomass and abundance have been reduced by fishing. And while the exact magnitude of depletion for particular stocks is often debatable, few would argue that there is no need for actions to protect against continued declines in some cases. Changes in size structure and genetic composition, localized depletions, and alterations in trophic structure of ecosystems are all occurring as well. However, effects of fishing are spatially heterogeneous and generalizations from one region or fishery to global fisheries can be misleading.

The evidence provided in Chapter 2 is compelling, but by no means exhaustive, concerning all possible effects. Ongoing research and assessment is required to understand the temporal and spatial extent of fishing impacts and how current and future management policies act to ameliorate or worsen the effects of fishing on marine food webs and species interactions.

Fisheries management strategies currently employed in the United States generally do not take into account ecosystem effects and multi-species interactions. Ecosystem considerations are discussed in regular stock assessments, but in general they do not involve a comprehensive evaluation of management strategies. Also, environmental impact statements are required for major fishery management actions, but there is no regulatory requirement to account for interactions among species. Instead, harvest policies tend to focus almost exclusively on single species, and maximum sustainable yield reference points are the norm.

In a multi-species context, interaction between species, mediated through predator-prey interactions and food-web effects, will need to be explicitly considered when deciding harvest strategies. For examples, if interacting species are both targeted for harvest, maximizing yield for each will likely be impossible. Tradeoffs between the allowable catch of each species will need to be made, and the desired level for each species' harvest decided simultaneously. This must occur while avoiding possible undesirable and irreversible shifts in the composition of the overall ecosystem. In addition, fisheries are not the only service that humans derive from the ocean, and incorporating these values into fisheries management decisions further increases the apparent number of tradeoffs.

Protecting ecosystem functioning and making allocation decisions between uses are two distinct—yet interdependent—issues that managers will have to face with increasing frequency as ecosystem considerations are factored into fisheries management. Within a functioning ecosystem, differing harvest and protection goals can be established amidst a variety of stewardship options and tradeoffs between uses. Greater understanding of these tradeoffs and the consequences of management actions are needed, but ultimately, society will need to decide the desired balance of services provided by the ocean and what protections to afford marine species. Based on these decisions, management approaches will need to be crafted to increase the chances that harvest controls are implemented effectively and in a manner that might further ecosystem considerations among users.

## RECOMMENDATIONS

The following recommendations fall into three categories: (1) implementing ecosystem considerations in fisheries management actions, (2) promoting stewardship, and (3) supporting future research.

### Applying a Food-Web Perspective to Fishery Management Strategies

**Multiple-species harvest strategies should be evaluated to account for species interactions and food-web dynamics.**

Setting multi-species harvest strategies requires taking into account food-web interactions, changes in trophic structure, life history strategies, and bycatch, all of which can change ecosystem productivity. If management is to account for the ecological interdependence among harvest targets and other food-web components, methods will be needed for quantitatively and qualitatively examining these interactions. Increased application of existing food-web, species-interaction, and ecosystem models, and development of new ones, can improve understanding of food-web effects and the impacts of fishing on ecosystem components, and also help to develop multi-species harvest strategies.

**Food-web, species-interaction, and ecosystem models should be used to evaluate alternative policy and management scenarios. These scenarios should elucidate the management tradeoffs that need to be made among user communities in a multi-species context and thereby inform the choice of multi-species harvest strategies.**

Fisheries management advice has tended to follow prescriptive policies defined in terms of generic biological reference points for individual populations as called for by the Magnuson-Stevens Fisheries Conservation and Management Act. However, within an ecosystem context, tradeoffs between conflicting management objectives should be made explicitly by evaluating consequences in terms of different measures of performance that reflect impacts of policy decisions on varied ecosystem components and uses. These tradeoffs will not only need to be made between competing fisheries, but should also account for the interactions of fisheries with other consumptive and nonconsumptive uses.

Although the availability of data required to build food-web models is generally limited, enough information exists for many systems to begin the development of new models now—these models should continue to improve based on new studies and information. The goal should not be to build a single best model, but to build a series of models as alternative hypotheses of what may happen, each representing a plausible scenario. Different fishery management strategies could be tested across the series of scenarios and the outcomes examined. Initially, analyses could be directed at near-term decision making by testing how current policies perform in an ecosystem context. Assigning relative likelihoods to each scenario will allow for tradeoffs between conflicting management objectives to be evaluated in order to better inform decision making. Therefore, model-based scenario analysis will encompass more than just determining reference points to be used as targets and limits. By choosing a particular management strategy, the feedback rules needed to calculate input- or output-based fishing controls will also be determined and the data needed to implement those rules will be specified.

Such approaches should be conceived as adaptive management experiments, with a requirement to implement monitoring programs to evaluate system responses and to detect unexpected consequences, should they happen. Scenario analysis will be best applied in an iterative manner and should continually incorporate new knowledge of the system as measured by outcomes of previous management actions (Figure 6.1). Model-based scenario analysis should first be applied in ecosystems that are relatively well known scientifically. Once a framework for scenario-based decision making has been established, scenario analysis can then be applied in ecosystems and fisheries that are potentially more contentious due to the tradeoffs that need to be made.

**Interdisciplinary working groups should be considered as a mechanism for developing appropriate models for each management area and for generating the series of scenarios needed to test proposed management actions.**

Building ecosystem models to design harvesting policies requires the cooperation of many specialists and the integration of information from many sources. To promote the development of ecosystem models for use in policy analysis, ecosystem-specific working groups could be established for particular areas of concern—composed of scientists and modelers drawn from the management agencies and academia, and representing both natural and social sciences.

Working groups would facilitate the consolidation of existing information, the generation of new syntheses with existing models, and the development of new models and other approaches to inform scenario development and forecasting under alternative management strategies. Working groups could meet with a variety of stakeholders—including fishermen and other consumptive and non-consumptive users—to identify important tradeoffs that should be considered when creating models and to evaluate feasible candidate policies. In particular, fishermen's historical knowledge of the resource and results of previous management actions may be an extremely valuable resource. The simulations created should be quantitative when possible, but even rigorous qualitative scenarios would be useful in some systems. Analyses and models generated by working groups should be made publicly available and be published in the peer-reviewed literature. Iterative analyses might be incorporated as the systems begin to respond to management actions.

**New governance and management instruments that create stewardship incentives among user groups should be evaluated and considered for adoption in the United States for multi-species fisheries management.**

Fisheries management has largely utilized a top-down system that places managers and users in an adversarial relationship that sometimes generates rash "race-to-fish" incentives among fishermen. But new and fundamentally different schemes adopted for some U.S. and international fisheries alter incentives and redirect fishing behavior in essential ways. All of these schemes use some form of dedicated access privilege, whereby consumptive users are granted secure shares in biologically determined allowable harvest targets. Previous reports indicate that secure access privileges provide new incentives to create value

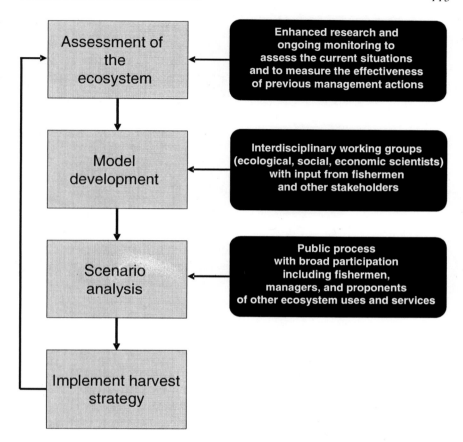

**FIGURE 6.1** The process of scenario analysis-based management should be an iterative and adaptive process. Improved data on food-web interactions, and changes in these interactions in both time and space, will help to create and update the models developed for a particular system. New and traditional regulatory schemes (catch and effort quotas set by different feedback control rules, marine protected areas, slot limits, gear type, etc.) and different monitoring schemes can, in principle, all be tested for their potential impacts on fished ecosystems and on user groups through the analysis process. Further, it is desirable that future models be set up to analyze the outcomes of different economic and social dynamics, behavior, and market pressures. Once there is a way to visualize all these different options, then a broad range of stakeholders can discuss which management schemes best achieve their collective goals and what tradeoffs are involved in deciding the management actions that should be taken. Monitoring and regular assessments will be needed to feed the management process and to determine how well the previous actions achieve the intended outcome, and data should be collected on how essential ecosystem components changed. This information will then feed back into model development, and a new round of evaluating of alternative management strategies would be initiated.

rather than to maximize harvest, and they may also generate a stakeholder interest in the long-term health of the resource.

Fisheries governance and ecosystem management options used by countries other than the United States and a few fisheries within the United States can be compared and examined for broader application in U.S. fisheries (nationwide). There are a number of specific governance options, each of which convey secure access privileges, including individual transferable quotas (ITQs), cooperative harvesting associations, and territorial harvesting institutions. Furthermore, many systems have begun to address ecosystem management issues such as bycatch reduction, habitat damages, marine reserve site selection, and nonconsumptive services. In some countries, there is a considerable history of experience that would allow for the assessment of strengths and weaknesses of these options in different biological and socioeconomic settings. There is also much to be learned from the few systems that have been established in the United States (i.e., sablefish/halibut individual fishing quota [IFQ] system), both from their successes and previous failures.

### Promoting Better Stewardship of the Marine Environment

**Fisheries management structures should ensure that a broad spectrum of social values is included in policy and management decisions.**

As previously recommended, ecosystem-based fisheries management approaches should be determined using model-based scenario analysis. However, this will not be a productive exercise if those affected by alternative outcomes are not involved in the decision-making process. Furthermore, proper stewardship of the marine environment requires consideration of values beyond just the commercial value of harvested species. Incorporating such values will require input both from fisheries scientists and social scientists—especially because fisheries management is about managing people and their behaviors as well as the biological resource itself.

To produce comprehensive management plans, fisheries managers should incorporate the best available social science as well as the best available natural science in their deliberations. The main objective should be to make better-informed decisions about tradeoffs across sectors and with other services. In other words, while natural science is essential for management, natural science by itself is not enough; proper management requires consideration for marine organisms as well as for human behavior and values. This will require greater involvement by social scientists with expertise in fields such as valuation, natural resource economics, decision science, and institutional design.

Governance structures should also be examined to ensure integral and effective representation of the public in deciding resource allocation tradeoffs. One of the principal goals of the Magnuson-Stevens Fishery Conservation and Management Act was to link the fishing community more directly to the management process. It has succeeded in that commercial and recreational fisheries are represented through the current organization of the Fisheries Management Councils, but nonconsumptive services and existence values are most certainly underrepresented. If model-based scenario analysis is used for making fisheries management decisions, nonconsumptive and public-good values will need to receive proper consideration when making tradeoffs among ocean services. Creating a mechanism for input for these other users could be accomplished at several stages of the management process, and NMFS and the Fishery Management Councils should continue to examine new avenues for greater stakeholder participation.

## Increasing Understanding of the Ecosystem Effects of Fishing

**Research is needed that improves understanding about the extent of fishing effects on marine ecosystems and promotes the development of ecosystem, food-web, and species-interaction models and their incorporation into management decisions.**

Data needs in support of ecosystem-based management will likely be more than the simple sum of currently available single-species information. Much more can be known about food-web linkages and interactions, including the strength of linkages between species and life-history stages and how these interactions change over time. In addition, modification of existing models and/or the development of new models are needed to better account for uncertainty in model output, to elucidate indicators of regime shifts and other interacting factors, and to evaluate monitoring schemes necessary to provide adequate information on ecosystem structure and function. To improve the utility of current models and their application to managed ecosystems, research should be conducted to provide a better understanding of:

- the dynamics of food-web interactions, including food habit data and interactions at lower trophic levels,
- per capita effects or population effects so that dynamical changes at a variety of trophic levels can be evaluated,
- whether models possibly ignore species with strong interactions in the ecosystem, and
- the need for baseline data on a number of non-target and lower-trophic species to determine the role of these species in the ecosystem when their abundances and interactions change due to fishing pressure.

Spatially explicit biological data should be collected that will allow both large-scale population trends and changes at finer scales to be monitored and understood. Patterns of interaction and the strength of these interactions vary in time and space. Collecting data on both temporal and spatial scales will allow for the variability in these interactions in both dimensions to be examined, thus improving the use of models to generate future scenarios. This may also lead to management approaches that can take into account temporal and spatial fluctuations in species interactions, biomass, and life history.

Research is also needed to determine whether interactions at various scales can help to define ecosystem boundaries in a new way. Biologically relevant boundaries in the marine environment are virtually impossible to identify, making it difficult to set the boundaries for modeling and management. Examining population trends and species interactions on finer spatial scales may help to define important biological interactions, which can then be bounded by management goals, financial constraints, and other realities imposed by political considerations.

Historical data assessments will be necessary to provide new insights about past species abundances and interactions. Comprehensive analyses of existing data can be applied to ongoing changes in target species, and they can help identify changes in habitat and non-target species that are thought to indicate ecosystem status. Landings data, narratives and descriptions, fisheries-independent data, phytoplankton and plankton records, satellite data, and archived specimens should all be considered when conducting these types of analyses. Examining these time-series or snapshot data in ecosystem and food-web models may provide the best approach to synthesizing long-term data and identifying alternative future scenarios to evaluate policy choices. Determining historical levels of exploited populations, and their natural fluctuations, should also provide a baseline around which to establish future management actions, including the setting of recovery goals.

**Research is needed that expands relevant social and economic information and the integration of this knowledge in fisheries management actions.**

Future research should measure and evaluate nonconsumptive uses of marine ecosystem services and quantify the non-use values of marine ecosystem components. Research is also needed to determine if (and how) social and economic principles vary from ecosystem to ecosystem. Concepts from decision science and financial portfolio theory are well developed and can be used to inform decisions in other settings such as in discussions of multi-species ecosyste objectives.

Understanding the social and ethical values associated with the broad suite of services provided by marine ecosystems is important and will require measurement and scaling of those values in relation to other uses. While economists and cognitive psychologists have examined these issues for terrestrial systems, there is little comparable work for marine systems. This too will require collaborative research that brings marine scientists together with social scientists. The task for marine scientists is to elucidate how various fishing strategies affect the structure of marine ecosystems and alter fundamental processes operating in marine systems. Valuation researchers can then design methods to evaluate how various changes affect humans directly and indirectly and assess those changes to uncover policy options that reflect the most desirable choices. Furthermore, integrated biologic-socioeconomic models should be explored for their capacity to capture important biophysical linkages, translated through to human impacts via economic market valuation methods.

Evaluating management options will require integrated modeling that incorporates not only the best depictions of ecosystem links, but also accurate depictions of fishermen's behaviors and responsiveness to changes in governance systems. Understanding behavior is a particularly under-researched area, even behavior associated with conventional management systems. Part of the problem is that the data collected in most fisheries are designed by managers for managing with conventional top-down approaches. These data often are aggregated in ways that obscure critical information about the microbehavior of fishermen that would be useful for forecasting behavior under existing or alternative systems. Information should be collected to examine how different kinds of governance mechanisms could potentially change fishermen's behavior rather than simply regulate it.

Research should also be conducted on how ecosystem management objectives can be incorporated into incentive-based governance mechanisms. Most existing incentive-based systems are primarily single-species focused, but many are also beginning to address broader ecosystem objectives. Existing experiences should be examined and new research conducted on management structures that might best address interspecies linkages, bycatch questions, and broader portfolios of ecosystem services. Research is needed to examine outcomes that might emerge by allowing various user groups to trade allocations. Similarly, a range of experience exists about how bycatch and discards can be incorporated within traditional ITQ or cooperative systems.

**New data management, archiving, and access methods should be developed for fisheries research and management.**

Fisheries data currently are fragmented and dispersed, thus slowing the use of these data in comprehensive analyses. Improvements to information systems would increase access to historical data, incorporate data from disparate sources, and support the policy-making process. Access to data from diverse sources will facilitate the transformation of these data into useful information that leads to model-based scenario analysis and informed decisions. Large-scale modeling of marine ecosystems requires facile integration of data from multiple sources. Better data management is fundamental to implementing ecosystem-based management of fisheries.

At the outset, the fisheries management community needs to examine existing approaches for the collection and standardization of ecosystem-level data, including those required by programs, such as the National Science Foundation's Long-Term Ecological Research, to facilitate across-system comparisons. There is an additional need for a repository and data management system for ecosystem-level research that will allow access to data through multiple-user portals. A comprehensive spatial database of fishing effort, harvest, and other relevant factors could be established, integrating new information management technologies.

Development of an information technology infrastructure that would provide a highly functional platform for scientific endeavors requires meeting challenges in a number of areas: data discovery, access, evaluation, and integration. Other desirable elements include a framework for modeling that provides repositories and documentation for models and model output, frameworks for designing and executing complex workflows, and advanced analysis and visualization tools. Furthermore, the technologic capabilities for ecosystem modeling will be quite intensive. The necessary data management and technology infrastructure should be an essential component of model analysis and scenario generation.

# References

Aldebert, Y. 1997. Demersal resources of the Gulf of Lyons. Impact of exploitation on fish diversity. *Vie Milieu* 47(4):275–284.

Aydin, K., and P. Livingston. 2003. *Food Web Comparisons in the Eastern and Western Bering Sea.* In: *AFSC Quarterly Report.* U.S. Department of Commerce, National Oceanic and Atmospheric Administration, Alaska Fisheries Science Center, Seattle, WA.

Baker, C.S., and P.J. Clapham. 2004. Modeling the past and future of whales and whaling. *Trends in Ecology and Evolution* 19:365.

Barange, M., F. Werner, I. Perry, and M. Fogarty. 2004. The tangled web: Global fishing, global climate, and fish stock fluctuations. *GLOBEC Newsletter* 10(2):20–23.

Barbier, E. 1994. Valuing environmental functions: Tropical wetlands. *Land Economics* 70(2):155–173.

Bascompte, J., C.J. Melián, and E. Sala. 2005. Interaction strength combinations and the overfishing of a marine food web. *Proceedings of the National Academy of Sciences* [USA] 102(15):5443–5447.

Baumgartner, T.R., A. Soutar, and V. Ferreira-Bartrina. 1992. Reconstruction of the history of Pacific sardine and northern anchovy populations over the past two millennia from sediments of the Santa Barbara Basin, California. *California Cooperative Oceanic Fisheries Investigations Report* 33:24–40.

Beamish, R.J., ed. 1995. *Climate Change and Northern Fish Populations.* Canadian Special Publication of Fisheries and Aquatic Sciences. Vol. 121. National Research Council, Ottawa, Canada, 739 pp.

Beamish, R.J., A.J. Benson, R.M. Sweeting, and C.M. Neville. 2004. Regimes and the history of the major fisheries off Canada's west coast. *Progress in Oceanography* 60:355–385.

Beddington, J.R., and R.M. May. 1977. Harvesting natural populations in a randomly fluctuating environment. *Science* 197:463–465.

Beisner, B.E., D.T. Haydon, and K. Cuddington. 2003. Alternative stable states in ecology. *Frontiers in Ecology and the Environment* 1:376–382.

Benson, A.J., and A.W. Trites. 2002. Ecological effects of regime shifts in the Bering Sea and eastern North Pacific Ocean. *Fish and Fisheries* 3:95–113.

Berkeley, S.A., C. Chapman, and S.M. Sogard. 2004. Maternal age as a determinant of larval growth and survival in marine fish, *Sebastes melanops*. *Ecology* 85:1258–1264.

Birkeland, C. 1997. *Life and Death of Coral Reefs*. Chapman & Hall, New York, NY.

Bjorndal, T., and J. Conrad, 1987. The dynamics of an open access fishery. *Canadian Journal of Economics* 20:74–85.

Bjorndal, K.A., and J.B.C. Jackson. 2003. Roles of sea turtles in marine ecosystems: Reconstructing the past. In: *The Biology of Sea Turtles*, Vol. II, P.L. Lutz, J.A. Musick and J. Wyneken, eds. CRC Press, Boca Raton, FL, 259–274.

Bockstael, N.E., and J. Opaluch. 1983. Discrete modeling of supply response under uncertainty: The case of the fishery. *Journal of Environmental Economics and Management* 10:125–137.

Botsford, L.W., J.C. Castilla, and C.H. Peterson. 1997. The management of fisheries and marine ecosystems. *Science* 277:509–515.

Breitburg, D.L., and G.F. Reidel. 2005. Multiple stressors in marine systems. In: *Marine Conservation Biology: The Science of Sustaining the Sea's Biodiversity*, E.A. Norse, and L.B. Crowder, eds. Island Press, Washington, DC, 167–182.

Bundy, A. 2005. Structure and functioning of the eastern Scotian Shelf ecosystem before and after the collapse of groundfish stocks in the early 1990s. *Canadian Journal of Fisheries and Aquatic Science* 62:1453–1473.

Bundy, A., and L.P. Fanning. 2005. Can Atlantic cod (*Gadus morhua*) recover? Exploring trophic explanations for the non-recovery of the cod stock on the eastern Scotian Shelf, Canada. *Canadian Journal of Fisheries and Aquatic Science* 62:1474–1489.

Butterworth, D.S., and A.E. Punt. 1999. Experiences in the evaluation and implementation of management procedures. *ICES Journal of Marine Science* 56:985–998.

Caddy, J.F., and S.M. Garcia. 1986. Fisheries thematic mapping. A prerequisite for intelligent management and development of fisheries. *Océanographie Tropicale* 21(1):31–52.

Caddy, J.F., J. Csirke, S.M. Garcia, and R.J.R. Grainger. 1998. How pervasive is "fishing down marine food webs"? *Science* 282:1383a.

Callicott, J.B. 1995a. Conservation Ethics at the Crossroads. In: *Evolution and the Aquatic Ecosystem: Defining Unique Units in Population Conservation*, J.L. Nielsen, and D.A. Bowers, eds. American Fisheries Society Symposium, Vol. 17. American Fisheries Society, Bethesda, MD, 3–7.

Callicott, J.B. 1995b. Conservation ethics and fishery management. *Fisheries* 16(2):22–28.

Carlton, J.T., J.B. Geller, M.L. Reaka-Kudla, and E.A. Norse. 1999. Historical extinctions in the sea. *Annual Review of Ecology and Systematics* 30:525–538.

Carpenter, S.R. 2003. *Regime Shifts in Lakes and Ecosystems: Patterns and Variation*. Publication of Excellence in Ecology. Book 15 (Series). O. Kinne, ed. Ecological Institution, Oldendorf/Luhe, Germany.

Carpenter, S.R., and W.A. Brock. 2004. Spatial complexity, resilience and policy diversity: Fishing on lake-rich landscapes. *Ecology and Society* 9(1):8.

Carpenter, S.R., J.F. Kitchell, and J.R. Hodson. 1985. Cascading trophic interactions and lake productivity. *BioScience* 35:634–639.

Carpenter, S.R., T.M. Frost, J.F. Kitchell, T.K. Kratz, D.W. Schindler, J. Shearer, W.G. Sprules, M.J. Vanni, and A.P. Zimmerman. 1991. Patterns of primary production and herbivory in 25 North American lake ecosystems. In: *Comparative Analyses of Ecosystems, Patterns, Mechanisms, and Theories*, J. Cole, G. Lovett, and S. Findley, eds. Springer-Verlag, New York, NY, 96–97.

Carpenter, S., W. Brock, and P. Hanson. 1999. Ecological and social dynamics in simple models of ecosystem management. *Conservation Ecology* 3(2):4.

CCSBT (Commission for the Conservation of Southern Bluefin Tuna). 2005. *Report of the Fourth Meeting of the Management Procedure Workshop, Canberra, Australia*. Commission for the Conservation of Southern Bluefin Tuna, Canberra, Australia.

Christensen, V. 1998. Fishery-induced changes in a marine ecosystem: Insight from models of the Gulf of Thailand. *Journal of Fish Biology* 53(Suppl. A):128–142.
Christensen, V. 2005. *Using Ecosystem Modeling to Evaluate Ecosystem Effects of Fishing: Status and Outlook.* Presentation to the National Research Council Committee on Ecosystem Effects of Fishing: Phase II—Assessments of the Extent of Change and the Implications for Policy, Meeting 2, held on May 9, 2005, in Seattle, WA. National Research Council, Ocean Studies Board, Washington, DC.
Christensen, V., and D. Pauly. 1992. ECOPATH II—A software for balancing steady-state ecosystem models and calculating network characteristics. *Ecological Modeling* 61:169–185.
Christensen, V., and D. Pauly. 2004. Placing fisheries in their ecosystem context, an introduction. *Ecological Modeling* 172:103–107 (editorial).
Christensen, V., and C. Walters. 2004. Trade-offs in ecosystem-scale optimization of fisheries management policies. *Bulletin of Marine Science* 74(3):549–562.
Christensen, V., J.E. Beyer, H. Gislason, and M. Vinter. 2002. A comparative analysis of the North Sea based on Ecopath with Ecosim and multi-species virtual population analysis. In: *Proceedings of the INCO-DC Conference Placing Fisheries in their Ecosystem Context*, Galápagos Islands, Ecuador, December 4–8, 2000, 48.
Christensen, V., S. Guenette, J.J. Heymans, C.J. Walters, R. Watson, D. Zeller, and D. Pauly. 2003. Hundred-year decline of North Atlantic predatory fishes. *Fish and Fisheries* 4:1–24.
Christensen, V., P. Amorium, I. Diallo, T. Diouf, S. Guenete, J.J. Heymans, A.N. Mendy, T. Mahfoudh Sidi, M.L.D. Palomares, B. Samb, K. Stobberup, J.M. Vakily, M. Vasconcellos, R. Watson, and D. Pauly. 2004. Trends in fish biomass off Northwest Africa, 1960–2000. In: *West African Marine Ecosystems: Models and Fisheries Impacts*, M.L.D. Palomares and D. Pauly, eds. Publication of Fisheries Centre Research Reports 12(7). Fisheries Centre, University of British Columbia, Vancouver, Canada, 215–220.
Chuenpagdee, R., L.E. Morgan, S.M. Maxwell, E.A. Norse, and D. Pauly. 2003. Shifting gears: Assessing collateral impacts of fishing methods in U.S. waters. *Frontiers in Ecology and the Environment* 1(10):517–524.
CITES (Convention on International Trade in Endangered Species). 2005. [Online] Available at: http://www.cites.org [accessed April 19, 2006].
Clark, C. 1985. *Bioeconomic Modeling and Fisheries Management.* John Wiley and Sons, New York, NY.
Clark, J.S., S.R. Carpenter, M. Barber, S. Collins, A. Dobson, J. Foley, D. Lodge, M. Pascual, R. Pielke Jr., W. Pizer, C. Pringle, W. Reid, K. Rose, O. Sala, W. Schlesinger, D. Wall, and D. Wear. 2001. Ecological forecasts: An emerging imperative. *Science* 293:657–660.
Cole, J., and J. McGlade. 1998. Clupeoid population variability, the environment and satellite imagery in coastal upwelling systems. *Reviews in Fish Biology and Fisheries* 8:445–471.
Coleman, F.C., C.C. Koenig, A.M. Eklund, and C.B. Grimes. 1999. Management and conservation of temperate reef fishes in the grouper-snapper complex of the southeastern United States. In: *Life in the Slow Lane: Ecology and Conservation of Long-Lived Marine Animals*, J.A. Musick, ed. American Fisheries Society Symposium 23. American Fisheries Society, Bethesda, MD, 233–242.
Collie, J.S., and H. Gislason. 2001. Biological reference points for fish stocks in a multispecies context. *Canadian Journal of Fisheries and Aquatic Sciences* 58:2233–2246.
Collie, J.S., H. Gislason, and M. Vinther. 2003. Using AMOEBAs to display multispecies, multifleet fisheries advice. *ICES Journal of Marine Science* 60:709–720.
Collie, J.S., K. Richardson, and J.H. Steele. 2004. Regime shifts: Can ecological theory illuminate the mechanisms? *Progress in Oceanography* 60:281–302.
Connell, J.H., and W.P. Sousa. 1983. On the evidence needed to judge ecological stability or persistence. *American Naturalist* 121:789–824.

Conover, D.O., and S.B. Munch. 2002. Sustaining fisheries yields over evolutionary time scales. *Science* 297:94–96.

Cox, S.P., S.J.D. Martell, C.J. Walters, T.E. Essington, J.F. Kitchell, C. Boggs, and I. Kaplan. 2002a. Reconstructing ecosystem dynamics in the central Pacific Ocean, 1952–1998. I. Estimating population biomass and recruitment of tunas and billfishes. *Canadian Journal of Fisheries and Aquatic Sciences* 59:1724–1735.

Cox, S.P., T.E. Essington, J.F. Kitchell, S.J.D. Martell, C.J. Walters, C. Boggs, and I. Kaplan. 2002b. Reconstructing ecosystem dynamics in the central Pacific Ocean, 1952–1998. II. A preliminary assessment of the trophic impacts of fishing and effects on tuna dynamics. *Canadian Journal of Fisheries and Aquatic Sciences* 59:1736–1747.

Craig, J.K., L.B. Crowder, C.D. Gray, C.J. McDaniel, T.A. Henwood and J.G. Hanifen. 2001. Ecological effects of hypoxia on fish, sea turtles and marine mammals in the northwestern Gulf of Mexico. In: *Coastal Hypoxia: Consequences for Living Resources and Ecosystems*, N. Rabalais and R.E. Turner, eds. Publication of Coastal and Estuarine Studies 58. American Geophysical Union, Washington, DC, 269–291.

Croll, D.A., J.L. Maron, J.A. Estes, E.M. Danner, and G.V. Byrd. 2005. Introduced predators transform subarctic islands from grassland to tundra. *Science* 307:1959–1961.

Crowder, L.B., D. Reagan, and D.W. Freckman. 1996. Food-web dynamics and applied problems. In: *Food Webs: Integration of Patterns and Dynamics*, G.A. Polis and K.O. Winemiller, eds. Chapman & Hall, New York, NY, 327–336.

D'Agrosa, C., C.E. Lennert Cody, and O. Vidal. 2000. Vaquita bycatch in Mexico's gillnet fisheries: Driving a small population to extinction. *Conservation Biology* 14:1110–1119.

Day, D. 1989. *Vanished Species*. London Editions, London, United Kingdom.

Dayton, P.K., S.F. Thrush, M.T. Agardy, and R.J. Hofman. 1995. Environmental effects of marine fishing. *Aquatic Conservation: Marine and Freshwater Ecosystems* 5:205–232.

Dayton, P.K., M.J. Tegner, P.B. Edwards, and K.L. Riser. 1998. Sliding baselines, ghosts, and reduced expectations in kelp forest communities. *Ecological Applications* 8:309–322.

DiNardo, G.T., and J.A. Wetherall. 1999. Accounting for uncertainty in the development of harvest strategies for the Northwestern Hawaiian Islands lobster trap fishery. *ICES Journal of Marine Science* 56:943–951.

Dulvy, N.K., Y. Sadovy, and D. Reynolds. 2003. Extinction vulnerability in marine populations. *Fish and Fisheries* 4:25–64.

Eales, J., and J. Wilen. 1985. An examination of fishing location choice in the pink shrimp fishery. *Marine Resource Economics* 2:331–351.

Edmunds, P.J., and R.C. Carpenter. 2001. Recovery of *Diadema antillarum* reduces macroalgal cover and increases abundance of juvenile corals on a Caribbean reef. *Proceedings of the National Academy of Sciences* [USA] 98(9):5067–5071.

Elton, C.S. 1927. *Animal Ecology*. Sidgwick and Jackson, London, United Kingdom.

Essington, T. 2004. Getting the right answer from the wrong model: Evaluating the sensitivity of multi-species fisheries advice to uncertain species interactions. *Bulletin of Marine Science* 74(3):563–581.

Essington, T.E., and S. Hansson. 2004. Predator-dependent functional responses and interaction strengths in a natural food web. *Canadian Journal of Fisheries and Aquatic Sciences* 61:2227–2236.

Essington, T.E., A.H. Beaudreau, and J. Wiedenmann. 2006. Fishing through marine food webs. *Proceedings of the National Academy of Sciences* [USA] 103:3171–3175.

Estes, J.A., and D.O. Duggins. 1995. Sea otters and kelp forests in Alaska: Generality and variation in a community ecological paradigm. *Ecological Monographs* 65:75–100.

Estes, J.A., M.T. Tinker, T.M. Williams, and D.F. Doak. 1998. Killer whales predation on sea otters linking oceanic and nearshore ecosystems. *Science* 282:473–476.

# REFERENCES

Estes, J.A., M.T. Tinker, A.M. Doroff, and D.M. Burns. 2005. Continuing sea otter population declines in the Aleutian archipelago. *Marine Mammal Science* 21:169–172.

FAO (Food and Agriculture Organization). 1996. *Precautionary Approach to Capture Fisheries and Species Introductions.* FAO Technical Guidelines for Responsible Fisheries. No 2. ISSN 1020-5292. Food and Agriculture Organization, Rome, Italy.

FAO. 2002. *The State of World Fisheries and Aquaculture.* Food and Agriculture Organization of the United Nations, Rome, Italy.

FAO. 2005. *Review of the State of World Marine Fishery Resources.* FAO Fisheries Technical Paper 457. Food and Agriculture Organization of the United Nations, Rome, Italy.

Ferraro, P., and A. Kiss. 2003. Direct payments to conserve biodiversity. *Science* 298:1718–1719.

Finney, B.P., I. Gregory-Eaves, M.S.V. Douglas, and J.P. Smol. 2002. Fisheries productivity in the northeastern Pacific Ocean over the past 2200 years. *Nature* 416:729–733.

Fogarty, M.J., and S.A. Murawski. 1998. Large-scale disturbances and the structure of marine systems: Fishery impacts on Georges Bank. *Ecological Applications* 8:S6–S22.

Francis, G.R., J.J. Magnuson, H.A. Regier, and D.R. Talhelm, eds. 1979. *Rehabilitating Great Lakes Ecosystems.* Technical Report No. 37. Great Lakes Fisheries Commission, Ann Arbor, MI.

Francis, R.C., S.R. Hare, A.B. Hollowed, and W.S. Wooster. 1998. Effects of interdecadal climate variability on the oceanic ecosystems of the NE Pacific. *Fisheries Oceanography* 7(1):1–21.

Frank, K.T., B. Petrie, J.S. Choi, and W.C. Leggett. 2005. Trophic cascades in a formerly cod-dominated ecosystem. *Science* 308:1621–1623.

Freeman, A.M. 1993. *The Measurement of Environmental and Resource Values: Theory and Methods.* Resources for the Future, Washington, DC.

Fulton, E.A., M. Fuller, A.D.M. Smith, and A. Punt. 2004. *Ecological Indicators of the Ecosystem Effects of Fishing: Final Report.* Report Number R99/1546. Commonwealth Scientific and Industrial Research Organisation, Aspendale, Australia.

Garcia, S.M., and M. Hayashi. 2000. Division of the oceans and ecosystem management: A contrastive spatial evolution of marine fisheries governance. *Ocean and Coastal Management* 43:445–474.

Garcia, S.M., and R.J.R. Grainger. 2005. Gloom and doom? The future of marine capture fisheries. *Philosophical Transactions of the Royal Society B-Biological Sciences* 360:21–46.

Gause, G.F. 1934. *The Struggle for Existence.* Williams and Wilkins, Baltimore, MD.

Golley, F.B. 1993. *A History of the Ecosystem Concept in Ecology: More Than the Sum of the Parts.* Yale University Press, New Haven, CT.

Grafton, R.Q., L.H. Pendelton, and H.W. Nelson. 2001. *A Dictionary of Environmental Economics, Science, and Policy.* Eldward Elgar, Northhampton, MA.

Gulland, J.A., ed. 1988. *Fish Population Dynamics: The Implications for Management.* 2nd ed. John Wiley and Sons, Chichester, United Kingdom.

Gulland, J.A., and S.M. Garcia. 1984. Observed patterns in multispecies fisheries. In: *Exploitation of marine communities*, R.M. May, ed. Springer Verlag, New York, NY, 155–190.

Hampton, J., J.R. Sibert, P. Kleiber, M.N. Maunder, and S. Harley. 2005. Decline of pacific tuna populations exaggerated? *Nature* 434:E1–E2.

Hardy, A.C. 1924. The herring in relation to its animate environment. Part I. The food and feeding habits of the herring with special reference to the east coast of England. *Fisheries Investigations* 7(3–Series II):53.

Hare, S.R., and N.J. Mantua. 2000. Empirical evidence for North Pacific regime shifts in 1977 and 1989. *Progress in Oceanography* 47(2–4):103–145.

Hargrove, E. 1995. The state of environmental ethics. In: *Environmental Ethics and the Global Marketplace*, D.G. Dallmeyer and A.F. Ike, eds. University of Georgia Press, Athens, GA, 16–30.

Heino, M., U. Dieckmann, and O.R. Godo. 2002. Estimating reaction norms for age and size at maturation with reconstructed immature size distributions: A new technique illustrated by application to Northeast Arctic cod. *ICES Journal of Marine Science* 59:562–575.

Heppell, S.S., S.A. Heppell, A.J. Read, and L.B. Crowder. 2005. Effects of fishing on long-lived marine organisms. In: *Marine Conservation Biology: The Science of Maintaining the Sea's Biodiversity*, E.A. Norse, and L.B. Crowder, eds. Island Press, Washington, DC, 211–231.

Hilborn, R., and M. Ledbetter. 1979. Analysis of the British Columbia salmon purse seine fleet: Dynamics of movement. *Journal of the Fisheries Research Board of Canada* 36:384–391.

Hilborn, R., and C. Walters. 1992. *Quantitative Fisheries Stock Assessment: Choice, Dynamics, and Uncertainty*. Chapman and Hall Inc., New York, NY.

Hilborn, R., T.A. Branch, B. Ernst, A. Magnussson, C.V. Minte-Vera, M.D. Scheuerell, and J.L. Valero. 2003. State of the world's fisheries. *Annual Review of Environment and Resources* 28:359–399.

Hilborn, R., K. Stokes, J. Maguire, T. Smith, L.W. Botsford, M. Mangel, J. Orensanz, A. Parma, J. Rice, J. Bell, K.L. Cochrane, S. Garcia, S.J. Hall, G.P. Kirkwood, K. Sainsbury, G. Stefansson, and C. Walters. 2004. When can marine reserves improve fisheries management? *Ocean and Coastal Management* 47:197–205.

Hinke, J.T., I.C. Kaplan, K. Aydin, G.M. Watters, R.J. Olson, and J.F. Kitchell. 2004. Visualizing the food-web effects of fishing for tunas in the Pacific Ocean. *Ecology and Society* 9(1):10.

Holland, D. 2000. A bioeconomic model of marine sanctuaries on Georges Bank. *Canadian Journal of Fisheries and Aquatic Sciences* 57(6):1307–1319.

Holling, C.S. 1973. Resilience and stability of ecological systems. *Annual Review of Ecology and Systematics* 4:1–23.

Hollowed, A.B., and W.S. Wooster. 1992. Variability of winter ocean conditions and strong year classes of Northeast Pacific groundfish. *ICES Marine Science Symposium* 195:433–444.

Hollowed, A.B., and W.S. Wooster. 1995. Decadal-scale variations in the eastern subarctic Pacific: II. Response of Northeast Pacific fish stocks. In: *Climate Change and Northern Fish Populations*, R.J. Beamish, ed. *Canadian Special Publication of Fisheries and Aquatic Sciences* 121:373–385.

Hollowed, A.B., N. Bax, R. Beamish, J. Collie, M. Fogarty, P. Livington, J. Pope, and J.C. Rice. 2000. Are multi-species models an improvement on single-species models for measuring fishing impacts on marine ecosystems? *ICES Journal of Marine Science* 57(3):707–719.

Hrbáček, J., M. Dvořáková, V. Kořínek and L. Procházková. 1961. Demonstration of the effect of the fish stock on the species composition of zooplankton and the intensity of metabolism of the whole plankton association. *Internationale Vereinigung für Theoretische und Angewandte Limnologie* 14:192–195.

Hseih, C., S.M. Glaser, A.J. Lucas, and G. Sugihara. 2005. Distinguishing random environmental fluctuations from ecological catastrophies for the North Pacific Ocean. *Nature* 435:336–340.

Hughes, T.P. 1994. Catastrophes, phase shifts, and large-scale degradation of a Caribbean coral reef. *Science* 256:1547–1551.

Hurrell, J.W., and H. van Loon. 1997. Decadal variations in climate associated with the North Atlantic Oscillation. *Climatic Change* 36:301–326.

Hutchings, J.A. 2000. Collapse and recovery of marine fishes. *Nature* 406:882–885.

Hutchings, J.A., and J.D. Reynolds. 2004. Marine fish population collapses: Consequences for recovery and extinction risk. *Bioscience* 54(4):297–309.

Hutchinson, G.E. 1959. Homage to Santa Rosalia, or why are there so many kinds of animals. *American Naturalist* 93:145–159.

Hyrenbach, K.D., K.A. Forney, and P.K. Dayton. 2000. Marine protected areas and ocean basin management. *Aquatic Conservation: Marine and Freshwater Ecosystems* 10:437–458.

IATTC (Inter-American Tropical Tuna Commission). 2004. Tunas and Billfishes in the Eastern Pacific Ocean in 2003. IATTC Fishery Status Report 2. Inter-American Tropical Tuna Commission, La Jolla, CA, 109 pp.

ICES (International Council for the Exploration of the Sea). 2000. Ecosystem effects of fishing. *ICES Journal of Marine Science* 57(3):2000.

# REFERENCES

IUCN (The World Conservation Union). 2001. *IUCN Red List Categories: Version 3.1*. IUCN Species Survival Commission, Gland, Switzerland.

IWC (International Whaling Commission). 2005. Report of the Scientific Committee, Annex D. Report of the Sub-Committee on the Revised Management Procedure. *Journal of Cetacean Research and Management* 7(Suppl.):84–92.

Jackson, J.B.C. 1997. Reefs since Columbus. *Coral Reefs* 16:S23.

Jackson, J.B.C., M.X. Kirby, W.H. Berger, K.A. Bjorndal, L.W. Botsford, B.J. Bourque, R.H. Bradbury, R. Cooke, J. Erlandson, J.A. Estes, T.P. Hughes, S. Kidwell, C.B. Lange, H.S. Lenihan, J.M. Pandolfi, C.H. Peterson, R.S. Steneck, M.J. Tegner, and R.R. Warner. 2001. Historical overfishing and the recent collapse of coastal ecosystems. *Science* 293:629–638.

Jennings, S., and J.L. Blanchard. 2004. Fish abundance with no fishing: Predictions based on macroecological theory. *Journal of Animal Ecology* 73:632–642.

Johnson, L.B., and S.H. Gage. 1997. Landscape approaches to the analysis of aquatic ecosystems. *Freshwater Biology* 37:113–132.

Jukic-Peladic, S., N. Vrgoc, S. Krstulovis-Sifner, C. Piccinette, G. Pacinetti-Manfrin, G. Marano, and N. Ungaro. 2001. Long-term changes in demersal resources of the Adriatic Sea: Comparison between trawl surveys carried out in 1948 and 1998. *Fisheries Research* 53:95–104.

Kareiva, P. 1994. Space: The final frontier in ecological theory. *Ecology* 75:1–47.

Kawasaki, T. 1992. Climate-dependent fluctuations in the Far Eastern sardine population and their impacts on fisheries and society. In: *Climate Variability, Climate Change, and Fisheries*, M.H. Glantz, ed. Cambridge University Press, Cambridge, 325–355.

Keleher, C.J., and F.J. Rahel. 1996. Thermal limits to salmonid distributions in the Rocky Mountain region and potential habitat loss due to global warming: A Geographic Information System (GIS) approach. *Transactions of the American Fisheries Society* 125:1–13.

Kellert, S.R. 2003. Human values, ethics, and the marine environment. In: *Values at Sea: Ethics for the Marine Environment*, D.G. Dallmeyer, ed. University of Georgia Press, Athens, GA, 1–18.

King, J., ed. 2005. Report of the Study Group on Fisheries and Ecosystem Responses to Recent Regime Shifts. PICES Scientific Report No. 28. North Pacific Marine Science Organization (PICES), Sidney, Canada.

Kitchell, J. 2005. *Lessons from ecosystem modeling*. Presentation to the National Research Council Committee on Ecosystem Effects of Fishing: Phase II—Assessments of the Extent of Change and the Implications for Policy, Meeting 1, held on March 29, 2005, Washington, DC. National Research Council, Ocean Studies Board, Washington, DC.

Kitchell, J.F., C. Boggs, X. He, and C.J. Walters. 1999. Keystone predators in the Central Pacific. In: *Ecosystem Approaches for Fisheries Management*, 16th Lowell Wakefield Fisheries Symposium. AL-SG-99-01. University of Alaska Sea Grant, Fairbanks, AK, 665–683.

Kitchell, J.F., I.C. Kaplan, S.P. Cox, S.J.D. Martell, T.E. Essington, C.H. Boggs, and C.J. Walters. 2004. Ecological and economic components of alternative fishing methods to reduce by-catch of marlin in a tropical pelagic ecosystem. *Bulletin of Marine Science* 74(3):607–619.

Knecht, R.W. 1992. Changing perceptions of the "American" coastal ocean. *Ocean and Coastal Management* 17:318–325.

Knowlton, N. 1992. Thresholds and multiple stable states in coral reef community dynamics. *American Zoologist* 32:674–682.

Koen-Alonso, M., and P. Yodzis. 2005. Multispecies modelling of some components of the marine community of northern and central Patagonia, Argentina. *Canadian Journal of Fisheries and Aquatic Sciences* 62:1490–1512.

Kraus, S.D., M. Brown, H. Caswell, C. Clark, M. Fujiwara, P. Hamilton, R. Kenney, A. Knowlton, S. Landry, C. Mayo, W. McLellan, M. Moore, D. Nowacek, D. Pabst, A.J. Read, and R. Rolland. 2005. North Atlantic right whales in crisis. *Science* 309:561–562.

Lange, G.M. 2003. *Policy Applications of Environmental Accounting*. World Bank, Washington DC.

Laska, M.S., and J.T. Wootton. 1998. Theoretical concepts and empirical approaches to measuring interaction strength. *Ecology* 79:461–476.

Law, R., and D.R. Grey. 1989. Evolution of yields from populations with age-specific cropping. *Evolution Ecology* 3:343–359.

Law, R., and K. Stokes. 2005. Evolutionary Impacts of fishing on target populations. In: *Marine Conservation Biology: The Science of Maintaining the Sea's Biodiversity*, E.A. Norse and L.B. Crowder, eds. Island Press, Washington, DC, 232–246.

Leopold, A. 1949. *A Sand County Almanac*. Oxford University Press, New York, NY.

Lessios, H.A. 1988. Mass mortality of *Diadema antillarum* in the Caribbean: What have we learned? *Annual Review in Ecological Systems* 19:371–393.

Levin, P. 2005. *EMOCC—An Ecosystem Model of the California Current*. Presentation to the National Research Council Committee on Ecosystem Effects of Fishing: Phase II—Assessments of the Extent of Change and the Implications for Policy, Meeting 3, held on June 30, 2005, in Washington, DC. National Research Council, Ocean Studies Board, Washington, DC.

Levin, S.A. 1992. The problem of pattern and scale in ecology. *Ecology* 73:1943–1967.

Lewison, R.L., and L. Crowder. 2003. Estimating fishery bycatch and effects on a vulnerable seabird population. *Ecological Applications* 13:743–753.

Lewison, R.L., L.B. Crowder, A.J. Read, and S.A. Freeman. 2004a. Understanding impacts of fisheries bycatch on marine megafauna. *Trends in Ecology and Evolution* 19:598–604.

Lewison, R.L., S.A. Freeman, and L.B. Crowder. 2004b. Quantifying the effects of fisheries on threatened species: The impact of pelagic longlines on loggerhead and leatherback sea turtles. *Ecology Letters* 7(3):221–232.

Lotka, A.J. 1925. *Elements of Physical Biology*. Williams & Wilkins Co., Baltimore, MD.

MacArthur, R.H. 1972. Strong, or weak, interactions? *Transactions of the Connecticut Academy of Arts and Sciences* 44:177–188.

MacCall, A.B. 2002. Fishery management and stock rebuilding prospects under conditions of low frequency variability and species interactions. *Bulletin of Marine Science* 70:613–628.

Mace, P.M. 2001. A new role for MSY in single species and ecosystem approaches to fisheries stock assessment and management. *Fish and Fisheries* 2:2–32.

Mace, P.M. 2004. In defense of fisheries scientists, single-species models and other scapegoats: Confronting the real problems. *Marine Ecology Progress Series* 274:285–291.

Magnuson, J.J. 1990. Long-term ecological research and the invisible present. *Bioscience* 40(7):495–501.

Magnuson, J.J., C.L. Harrington, D.J. Stewart, and G.N. Herbsat. 1981. Responses of macrofauna to short term dynamics of a Gulf Stream front on the Continental Shelf. In: *Coastal Upwelling (Coastal and Estuarine Sciences)*, R.A. Richards, ed. American Geophysical Union, Washington, DC, 441–448.

Mangel, M., and P.S. Levin. 2005. Regime, phase and paradigm shifts: Making community ecology the basic science for fisheries. *Philosophical Transactions of the Royal Society B-Biological Sciences* 360(1453):95–105.

Martell, S.J.D., A.I. Beattie, C.J. Walters, T. Nayar, and R. Briese. 2002. Simulating fisheries management strategies in the Strait of Georgia ecosystem using Ecopath and Ecosim. In: *The Use of Ecosystem Models to Investigate Multi-species Management Strategies for Capture Fisheries*, T. Pitcher and K. Cochrane, eds. Fisheries Centre Research Reports 10(2). University of British Columbia, Vancouver, Canada, 16–23.

Martinez, N.D. 1991. Artifacts or attributes of resolution on the Little Rock Lake food web. *Ecological Monographs* 61:367–392.

May, R.M., ed. 1984. *Exploitation of Marine Communities*. Springer-Verlag, New York, NY.

May, R.M., J.R. Beddington, C.W. Clark, S.J. Holt, and R.M. Laws. 1979. Management of multispecies fisheries. *Science* 203:267–277.

# REFERENCES

Menge, B.A. 1995. Indirect effects in marine rocky intertidal interaction webs: Patterns and importance. *Ecological Monographs* 65:21–74.

Millennium Assessment. 2003. *Ecosystems and Human Well-being: A Framework for Assessment*. Island Press, Chicago, IL.

Miller, A.J., and N. Schneider. 2000. Interdecadal climate regime dynamics in the North Pacific Ocean: Theories, observations, and ecosystem impacts. *Progress in Oceanography* 47:355–379.

MPA Center (National Marine Protected Areas Center). 2003a. *Regional Priorities for Social Science Research on Marine Protected Areas: U.S. Caribbean and South Florida*. Final Workshop Report. St. Croix, U.S. Virgin Islands, August 19–20, 2003.

MPA Center. 2003b. *Regional Priorities for Social Science Research on Marine Protected Areas: South Atlantic*. Final Workshop Report. Savannah, GA, December 2–4, 2003.

MPA Center. 2004. *Regional Priorities for Social Science Research on Marine Protected Areas: U.S. Pacific Islands*. Final Workshop Report. Waikoloa, HI, March 30–April 1, 2004.

Murdoch, W.W. 1969. Switching in general predators: Experiments on predator specificity and stability of prey populations. *Ecological Monographs* 39:335–354.

Myers, R.A., and B. Worm. 2003. Rapid worldwide depletion of predatory fish communities. *Nature* 423:280–283.

Myers, R.A., and B. Worm. 2005a. Reply to Hampton, J., J.R. Sibert, P. Kleiber, M.N. Maunder, and S. Harley. Decline of pacific tuna populations exaggerated? *Nature* 434:E1–E2.

Myers, R.A., and B. Worm. 2005b. Extinction, survival or recovery of large predatory fishes. *Philosophical Transactions of the Royal Society B-Biological Sciences* 360:13–20.

Myers, R.A., N. Barrowman, J. Hutchings, and A. Rosenberg. 1995. Population dynamics of exploited fish stocks at low population levels. *Science* 269:1106–1108.

Naeem, S., F.S. Chapin III, R. Costanza, P.R. Ehrlich, F.B. Golley, D.U. Hooper, J.H. Lawton, R.V. O'Neill, H.A. Mooney, O.E. Sala, A.J. Symstad, and D. Tilman. 1999. Biodiversity and ecosystem functioning: Maintaining natural life support processes. *Issues in Ecology* 4:2–11.

NMFS (National Marine Fisheries Service). 2005. *Annual Report to Congress on the Status of U.S. Fisheries—2004*. U.S. Department of Commerce, National Oceanic and Atmospheric Administration, Silver Spring, MD.

NOAA (National Oceanic and Atmospheric Administration). 2002. *Extension of Existing LOA under Section 120 of the Marine Mammal Protection Act*. National Oceanic and Atmospheric Administration, Northwest Regional Office, Seattle, WA.

Norse, E.A., L.B. Crowder, K. Gjerde, D. Hyrenbach, C.M. Roberts, C. Safina and M.E. Soule. 2005. Place-based ecosystem management in the open ocean. In: *Marine Conservation Biology: The Science of Maintaining the Sea's Biodiversity*, E.A. Norse, and L.B. Crowder, eds. Island Press, Washington, DC, 302–327.

Norton, B.G. 1986. On the inherent danger of undervaluing species. In: *The Preservation of Species: The Value of Biological Diversity*, B.G. Norton, ed. Princeton University Press, Princeton, NJ, 110–137.

Norton, B.G. 1991. *Toward Unity among Environmentalists*. Oxford University Press, New York, NY.

Norton, B.G. 2003. Marine environmental ethics: Where we might start. In: *Values at Sea: Ethics for the Marine Environment*, D.G. Dallmeyer, ed. University of Georgia Press, Athens, GA, 33–49.

NRC (National Research Council). 1996a. *The Bering Sea Ecosystem*. National Academy Press, Washington, DC.

NRC. 1996b. *Upstream: Salmon and Society in the Pacific Northwest*. National Academy Press, Washington, DC.

NRC. 1998. *Review of Northeast Fishery Stock Assessments*. National Academy Press, Washington, DC.

NRC. 1999a. *Sustaining Marine Fisheries*. National Academy Press, Washington, DC.

NRC. 1999b. *Sharing the Fish: Toward a National Policy on Individual Fishing Quotas*. National Academy Press, Washington, DC.

NRC. 2001. *Marine Protected Areas: Tools for Sustaining Ocean Ecosystems*. National Academy Press, Washington, DC.

NRC. 2002. *Effects of Trawling and Dredging on Seafloor Habitat*. National Academy Press, Washington, DC.

NRC. 2003. *Decline of the Stellar Sea Lion in Alaskan Waters: Untangling Food Webs and Fishing Nets*. The National Academies Press, Washington, DC.

NRC. 2004. *Valuing Ecosystem Services: Toward Better Environmental Decision-Making*. The National Academies Press, Washington, DC.

OITI Working Group. 2004. *Trends in Information Technology Infrastructure in the Ocean Sciences*. Ocean Information Technology Infrastructure Working Group, 24 pp. [Online] Available at: http://www.geo-prose.com/oceans_iti_trends [accessed April 19, 2006].

Olsen, E.M., M. Heino, G.R. Lilly, M.J. Morgan, J. Brattey, B. Ernande, and U. Dieckmann. 2004. Maturation trends indicative of rapid evolution preceded the collapse of northern cod. *Nature* 428:932–935.

Olson, R., and G. Watters. 2003. A model of the pelagic ecosystem in the eastern tropical Pacific Ocean. *Bulletin of the Inter-American Tropical Tuna Commission* 22(3):133–217.

Paine, R.T. 1966. Food web complexity and species diversity. *American Naturalist* 100:65–75.

Paine, R.T. 1969. A note on trophic complexity and community stability. *American Naturalist* 103:91–93.

Paine, R.T. 1980. Food webs: Linkage, interaction strength and community infrastructure. *Journal of Animal Ecology* 49:667–685.

Paine, R.T. 1992. Food-web analysis through field measurement of per capita interaction strength. *Nature* 355:73–75.

Palumbi, S.R., and J. Roman. 2004. Counting Whales in the North Atlantic—Response. *Science* 303:4.

Parma, A.M. 1990. Optimal harvesting of fish populations with non-stationary stock-recruitment relationships. *Natural Resource Modeling* 4:39–76.

Parma, A.M. 2002a. In search of robust harvest rules for Pacific halibut in the face of uncertain assessments and decadal changes in productivity. *Bulletin of Marine Science* 70:455–472.

Parma, A.M. 2002b. Bayesian approaches to the analysis of uncertainty in the stock assessment of Pacific halibut. In: *Incorporating Uncertainty into Fisheries Models*, J.M. Berkson, L.L. Kline and D.J. Orth, eds. American Fisheries Society, Bethesda, MD, 113–135.

Pauly, D. 1995. Anecdotes and the shifting baseline syndrome in fisheries. *Trends in Ecology and Evolution* 10:430.

Pauly, D., and V. Christensen. 1995. Primary production required to sustain global fisheries. *Nature* 374:255–257.

Pauly, D., and J.L. Maclean. 2003. *In a Perfect Ocean: The State of Fisheries and Ecosystems in the North Atlantic Ocean*. The State of the World's Oceans Series. Island Press, Washington, DC.

Pauly, D., and M.L. Palomares. 2005. Fishing down marine food webs: It is far more pervasive than we thought. *Bulletin of Marine Science* 76(2):197–211.

Pauly, D., V. Christensen, J. Dalsgaard, R. Froese, and F. Torres, Jr. 1998a. Fishing down marine food webs. *Science* 279:860–863.

Pauly, D., R. Froese, and V. Christensen. 1998b. How pervasive is "fishing down marine food webs": Response. *Science* 282:1383a.

Pauly, D., V. Christensen, and C. Walters. 2000. Ecopath, Ecosim, and Ecospace as tools for evaluating ecosystem impact of fisheries. *ICES Journal of Marine Science* 57:697–706.

Pauly, D., V. Christensen, S. Guenette, T.J. Pitcher, U.R. Sumaila, C.J. Walters, R. Watson, and D. Zeller. 2002. Towards sustainability in world fisheries. *Nature* 418:689–695.

Pauly, D., R. Watson, and J. Alder. 2005. Global trends in world fisheries: Impacts on marine ecosystems and food security. *Philosophical Transactions of the Royal Society B-Biological Sciences* 360:5–12.

# REFERENCES

Pearse, P., and J. Wilen. 1979. The impact of Canada's Pacific fleet salmon control program. *Journal of the Fisheries Research Board of Canada* 36(7):764–769.

Peterman, R.M., B.J. Pyper, and J.A. Grout. 2000. Comparison of parameter estimation methods for detecting climate-induced changes in productivity of Pacific salmon (*Onchorhynchus spp.*). *Canadian Journal of Fisheries and Aquatic Sciences* 57:181–191.

Peterson, C.H. 1984. Does a rigorous criterion for environmental identity preclude the existence of multiple stable points? *American Naturalist* 124:127–133.

Pew Oceans Commission. 2003. *America's Living Oceans: Charting a Course for Sea Change.* Pew Oceans Commission, Arlington, VA.

Pitcher, T. 2001. Fisheries managed to rebuild ecosystems: Reconstructing the past to salvage the future. *Ecological Applications* 11:601–617.

Plagányi, É.E., and D.S. Butterworth. 2004. A critical look at the potential of Ecopath with Ecosim to assist in practical fisheries management. *African Journal of Marine Science* 26:261–287.

Polacheck, T. In Press. Tuna longline catch rates in the Indian Ocean: Did industrial fishing result in a 90% decline in the abundance of large predatory species? *Marine Policy*.

Polis, G.A., and S.D. Hurd. 1996. Linking marine and terrestrial food webs: Allochthonous input from the ocean supports high secondary productivity on small islands and coastal land communities. *The American Naturalist* 147:396–423.

Polovina, J.J. 1984. Model of a coral reef ecosystem. I. The ECOPATH model and its application to French Frigate Shoals. *Coral Reefs* 3:1–11.

Polovina, J.J. 2005. Climate variation, regime shifts, and implications for sustainable fisheries. *Bulletin of Marine Science* 76:233–244.

Polovina, J.J., P. Kleiber, and D.R. Kobayashi. 1999. Application of *TOPEX-POSEIDON* satellite altimetry to simulate transport dynamics of larvae of spiny lobster, *Panulirus marginatus*, in the Northwestern Hawaiian Islands, 1993–1996. *Fishery Bulletin* 97(1):132–143.

Power, M.E. 1992. Top-down and bottom-up forces in food webs: Do plants have primacy? *Ecology* 73:733–746.

Punt, A.E., and D.S. Butterworth. 1995. The effects of future consumption by the Cape fur seal on catches and catch rates of the Cape hakes. 4. Modeling the biological interaction between Cape fur seals *Arctocephalus pusillus pusillus* and the Cape hakes *Merluccius capensis* and *M. paradoxus*. *South African Journal of Marine Science* 16:255–285.

Punt, A., and R. Hilborn. 2001. *BAYES-SA: Bayesian Stock Assessment Methods in Fisheries: User's Manual.* FAO Computerized Information Series (Fisheries). No 12. Food and Agriculture Organization of the United Nations, Rome, Italy.

Quero, J. 1998. Changes in the Euro-Atlantic fish species composition resulting from fishing and ocean warming. *Italian Journal of Zoology* 65:493–499.

Rabalais, N.M., R.E. Turner, and D. Scavia. 2002. Beyond science into policy: Gulf of Mexico hypoxia and the Mississippi River. *BioScience* 52:129–142.

Rawls, J. 1971. *A Theory of Justice.* Harvard University Press, Cambridge, MA.

Restrepo, V.R., P.M. Mace, and F. Serchuk. 1999. The precautionary approach: A new paradigm or business as usual? In: *Our Living Oceans: Report on the Status of U.S. Living Marine Resources.* NOAA Technical Memo. NMFS-F/SPO-41. U.S. Department of Commerce, National Marine Fisheries Service, Silver Spring, MD, 61–70.

Ricker, W.E. 1981. Changes in the average size and age of Pacific salmon. *Canadian Journal of Fisheries and Aquatic Sciences* 38:1636–1656.

Roman, J., and S.R. Palumbi. 2003. Whales before whaling in the North Atlantic. *Science* 301:508–510.

Rose, K.A., J.H. Cowan Jr., K.O. Winemiller, R.A. Myers, and R. Hilborn. 2001. Compensatory density dependence in fish populations: Importance, controversy, understanding and prognosis. *Fish and Fisheries* 2:293–327.

Rosenberg, A.A., W.J. Bolster, K.E. Alexander, W.B. Leavenworth, A.B. Cooper, and M.G. McKenzie. 2005. The history of ocean resources: Modeling cod biomass using historical records. *Frontiers in Ecology and the Environment* 3(2):84–90.

Rudd, M.A. 2004. An institutional framework for designing and monitoring ecosystem-based fisheries management policy experiments. *Ecological Economics* 48(1):109–124.

Russow, L-M. 1981. Why Do Species Matter? *Environmental Ethics* 3:101–112.

Ryther, J.H. 1969. Photosynthesis and fish production in the sea. *Science* 166:72–76.

Saenz-Arroyo, A., C.M. Roberts, J. Torre, M. Carino-Olvera, and R.R. Enriquez-Andrade. 2005. Rapidly shifting environmental baselines among fishers in the Gulf of California. *Proceedings of the Royal Society B* 272:1957–1962.

Safina, C. 2003. Launching a sea ethic. *Wild Earth* 12(4):2–5.

Sainsbury, K.J., A.E. Punt, and A.D.M. Smith. 2000. Design of operational management strategies for achieving fishery ecosystem objectives. *ICES Journal of Marine Science* 57:731–741.

Sala, E., O. Aburto-Oropeza, M. Reza, G. Paredes, and L.G. Lopez-Lemus. 2004. Fishing down coastal food webs in the Gulf of California. *Fisheries* 29(3):19–25.

Scammon, C.M. 1874. *The Marine Mammals of the Northwestern Coast of North America Together with an Account of the American Whale Fishery.* J.H. Carmany, San Francisco, CA.

Schindler, D.E., T.E. Essington, J.F. Kitchell, C. Boggs, and R. Hilborn. 2002. Sharks and tunas in the central Pacific: Fisheries impacts on predators with contrasting life histories. *Ecological Applications* 12:735–748.

Schoener, T.W. 1983. Field experiments on interspecific competition. *Theoretical Population Biology* 122:240–285.

Shannon, L.J., V. Christensen, and C.J. Walters. 2004. Modelling stock dynamics in the southern Benguela ecosystem for the period 1978–2002. *African Journal of Marine Science* 26:179–196.

Sheffer, M., S. Carpenter, J.A. Foley, and B. Walker. 2001. Catastrophic shifts in ecosystems. *Nature* 413:591–596.

Shelton, P.A., and B.P. Healey. 1999. Should depensation be dismissed as a possible explanation for the lack of recovery of the northern cod (*Gadus morhua*) stock? *Canadian Journal of Fisheries and Aquatic Sciences* 56:1521–1524.

Sherman, K. 1991. The large marine ecosystem concept—Research and management strategy for living marine resources. *Ecological Applications* 1(4):349–360.

Silliman, B.R., and M.D. Bertness. 2002. A trophic cascade regulates salt marsh primary production. *Proceedings of the National Academy of Sciences* [USA] 99(16):10500–10505.

Simenstad, C.A., J.A. Estes, and K.W. Kenyon. 1978. Aleuts, sea otters, and alternate stable-state communities. *Science* 200:403–411.

Sinclair, M. and G. Valdimarsson, eds. 2003. *Responsible Fisheries in the Marine Ecosystem.* Food and Agriculture Organization, Rome, Italy.

Smith, A.D.M. 2005. *EBFM—Aussie Style.* Presentation at the NOAA Workshop on Ecosystem-Based Decision Support Tools for Fisheries Management. Key Largo, FL, February 14–18. [Online] Available at: http://www.st.nmfs.noaa.gov/st7/ecosystem/workshop/2005/index.html [accessed April 19, 2006].

Smith, M.D., and J. Wilen. 2003a. Economic implications of marine reserves: The importance of spatial behavior. *Journal of Environmental Economics and Management* 46:183–206.

Smith, M.D., and J. Wilen. 2003b. Spatial Policy and Spatial Supply Response. *Marine Resource Economics* 19(1):85–112.

Smith, A.D.M., K.J. Sainsbury, and R.A. Stevens. 1999. Implementing effective fisheries management systems—Management strategy evaluation and the Australian partnership approach. *ICES Journal of Marine Science* 56:967–979.

Söderqvist, T., H. Eggert, B. Olsson, and A. Soutukorva. 2003. Economic valuation for sustainable development in the Swedish coastal zone. Working paper. Beijer International Institute for Ecological Economics of The Royal Swedish Academy of Sciences, Stockholm, Sweden.

# REFERENCES

Spencer, P.D. 1997. Optimal harvesting of fish populations with nonlinear rates of predation and autocorrelated environmental variability. *Canadian Journal of Fisheries and Aquatic Sciences* 54:59–74.

Sprague, L.M., and J.H. Arnold. 1972. Trends in the use and prospects for the future harvest of world fisheries resources. *Journal of the American Oil Chemists' Society* 49:345–350.

Springer, A.M., J.A. Estes, G.B. van Vliet, T.M. Williams, D.F. Doak, E.M. Danner, K.A. Forney, and B. Pfister. 2003. Sequential megafaunal collapse in the North Pacific Ocean: An ongoing legacy of industrial whaling? *Proceedings of the National Academy of Sciences* [USA] 100(21):12223–12228.

Steele, J.H. 1974. *The Structure of Marine Ecosystems*. Harvard University Press, Cambridge, MA.

Stefansson, G., and A.A. Rosenberg. 2005. Combining control measures for more effective management of fisheries under uncertainty: Quotas, effort limitation, and protected areas. *Philosophical Transactions of the Royal Society B-Biological Sciences* 360:133–146.

Steneck, R.S. 1998. Human influences on coastal ecosystems: Does overfishing create trophic cascades? *Trends in Ecology and Evolution* 13(11):429–430.

Sutherland, J.P. 1990. Perturbations, resistance, and alternative views of the existence of multiple stable points in nature. *American Naturalist* 136:270–275.

Swallow, S. 1994. Renewable and nonrenewable resource theory applied to coastal, agricultural, forest, wetland and fishery linkages. *Marine Resource Economics* 9:291–310.

Tam, W.M. 1992. Time, horizons, and the open sea. *Ocean and Coastal Management* 17:308–317.

Tilman, D. 1996. Biodiversity population versus ecosystem stability. *Ecology* 77(3):350–363.

Trippel, E.A., M.J. Morgan, A. Frechet, C. Rollet, A. Sinclair, C. Annand, D. Beanlands, and L. Brown. 1997. Changes in age and length at sexual maturity of northwest Atlantic cod, haddock and pollock stocks, 1972–1995. *Canadian Technical Report of Fisheries and Aquatic Sciences* 2157:xii.

Trites, A.W., P.A. Livingston, M.C. Vasconcellos, S. Mackinson, A. Springer, and D. Pauly. 1999. Ecosystem considerations and the limitations of Ecosim models in fisheries management: Insights from the Bering Sea. In: *Ecosystem Approaches for Fisheries Management*, S. Keller, ed. University of Alaska Sea Grant, Fairbanks, AK, 609–618.

Turner, M.G., W.H. Romme, and D.B. Tinker. 2003. Surprises and lessons from the 1988 Yellowstone fires. *Frontiers in Ecology and the Environment* 1(7):351–358.

U.S. Commission on Ocean Policy. 2004. *An Ocean Blueprint for the 21st Century*. U.S. Commission on Ocean Policy, Washington, DC.

UN (United Nations). 1995. *United Nations Conference on Straddling Fish Stocks and Highly Migratory Fish Stocks, Sixth Session*. July 24–August 4, 1995. United Nations, New York, N.Y.

van Kooten, G.C., and E.H. Bulte. 2000. *The Economics of Nature*. Blackwell Publishers Ltd., Oxford, United Kingdom.

Vitousek, P.M., H.A. Mooney, J. Lubchenco, and J.M. Melillo. 1997. Human domination of Earth's ecosystems. *Science* 227:494–499.

Volterra, V. 1926. Variations and fluctuations of the number of individuals in animal species living together. J. du Conseil intern. pour l'explor. de la mer III vol 1. Reprinted in *Animal Ecology*, 1931, by R.N. Chapman. McGraw-Hill, New York, NY.

Wahle, C., S.L.K. Barba, L. Bunce, P. Fricke, E. Nicholson, M. Orbach, C. Pomeroy, H. Recksiek, and J. Uravitch. 2003. *Social Science Research Strategy for Marine Protected Areas*. National Marine Protected Areas Center, MPA Science Institute, Santa Cruz, CA.

Walters, C. 2003. Folly and fantasy in the analysis of spatial catch rate data. *Canadian Journal of Fisheries and Aquatic Sciences* 60:1433–1436.

Walters, C., and A.M. Parma. 1996. Fixed exploitation rate strategies for coping with effects of climate change. *Canadian Journal of Fisheries and Aquatic Sciences* 53:148–158.

Walters, C., and J.F. Kitchell. 2001. Cultivation/depensation effects on juvenile survival and recruitment: Implications for the theory of fishing. *Canadian Journal of Fisheries and Aquatic Sciences* 58:1–12.

Walters, C., and S. Martell. 2004. *Fisheries Ecology and Management*. Princeton University Press, Princeton, NJ.

Walters, C., V. Christensen, and D. Pauly. 1997. Structuring dynamic models of exploited ecosystems from trophic mass-balance assessments. *Reviews in Fish Biology and Fisheries* 7:139–172.

Walters, C., D. Pauly, and V. Christensen. 1999. Ecospace: Prediction mesoscale spatial patterns in trophic relationships of exploited ecosystems, with emphasis on the impacts of marine protected areas. *Ecosystems* 2:539–554.

Walters, C., V. Christensen, S.J. Martell, and J.F. Mitchell. 2005. Possible ecosystem impacts of applying MSY policies from single-species assessments. *ICES Journal of Marine Science* 62:558–568.

Ward, P., and R.A. Myers. 2004. Shifts in open-ocean fish communities coinciding with the commencement of commercial fishing. *Ecology* 86(4):835–847.

Ware, D.M., and R.E. Thomson. 2005. Bottom-up ecosystem trophic dynamics determine fish production in the Northwest Pacific. *Science* 308:1280–1284.

Watson, R., and D. Pauly. 2001. Systematic distortions in world fisheries catch trends. *Nature* 414:536–538.

Wilen, J. 1976. *Open Access and the Dynamics of Overexploitation: The Case of the North Pacific Fur Seal*. UBC Natural Resource Economics Working Paper No. 3. University of British Columbia, Vancouver, Canada.

Wilen, J., M.D. Smith, D. Lockwood, and L. Botsford. 2002. Avoiding surprises: Incorporating fishermen behavior into management models. *Bulletin of Marine Science* 70:553–575.

Wiley, M.J., S.L. Kohler, and P.W. Seelbach. 1997. Reconciling landscape and local views of aquatic communities: Lessons from Michigan's trout streams. *Freshwater Biology* 37:133–148.

Winemiller, K.O., and K.A. Rose. 1992. Patterns of life-history diversification in North American fishes: Implications for population regulation. *Canadian Journal of Fisheries and Aquatic Sciences* 49:2196–2218.

Witman, J.D., and K.P. Sebens. 1992. Regional variation in fish predation intensity: A historical perspective in the Gulf of Maine. *Oecologia* 90:305–315.

Wootton, J.T. 1994. The nature and consequences of indirect effects. *Annual Review of Ecology and Systematics* 25:443–466.

Wootton, J.T. 1997. Estimates and tests of per capita interaction strength: Diet, abundance, and impact of intertidally foraging birds. *Ecological Monographs* 67(1):45–64.

Wootton, J.T., and M. Emmerson. 2005. Measurement of interaction strength in nature. *Annual Review of Ecology, Evolution, and Systematics* 36:419–444.

Worm, B., and J.E. Duffy. 2003. Biodiversity, productivity and stability in real food webs. *Trends in Ecology and Evolution* 18:628–632.

Worm, B., and R.A. Myers. 2003. Meta-analysis of cod–shrimp interactions reveals top-down control. *Nature* 423:280–283.

Worm, B., H.K. Lotze, C. Boström, R. Engkvist, V. Labanauskas, and U. Sommer. 1999. Marine diversity shift linked to interactions among grazers, nutrients and dormant propagules. *Marine Ecology Progress Series* 185:309–314.

Worm, B., M. Sandow, A. Oschlies, H.K. Lotze, and R.A. Myers. 2005. Global patterns of predator diversity in the open oceans. *Science* 306:1365–1369.

Yodzis, P. 1982. The compartmentalization of real and assembled food webs. *American Naturalist* 120:551–570.

# Appendix A

# Committee and Staff Biographies

## COMMITTEE MEMBERS

**John J. Magnuson** (*Chair*) is Professor Emeritus of Zoology and Limnology at the University of Wisconsin, Madison. He received his B.S. and M.S. degrees from the University of Minnesota and his Ph.D. degree in Zoology with a minor in Oceanography from the University of British Columbia. Dr. Magnuson was formerly Director of the University of Wisconsin's Center for Limnology and North Temperate Lakes Long-term Ecological Research Program. His research interests include long-term regional ecology, climate change effects on lake ecological systems, fish and fisheries ecology, and community ecology of lakes and islands. Previously a member of the Ocean Studies Board, Dr. Magnuson has a long and distinguished history of contributions to NRC projects including the OSB's Committee on Ecosystem Management for Sustainable Marine Fisheries, the Committee on Fisheries, and the U.S. National Scientific Committee on Oceanic Research.

**Dorinda G. Dallmeyer** is Associate Director of the Dean Rusk Center of International and Comparative Law at the University of Georgia. She holds three degrees from the University of Georgia: B.S. and M.S. degrees in Geology and a J.D. Prior to attending law school, Ms. Dallmeyer conducted research for over three years in tropical marine biology and ecology and collaborated on a number of scientific articles. Currently, Ms. Dallmeyer's primary research areas cross a broad spectrum of international law, with a particular emphasis on the role of negotiation and dispute resolution. She served as a principal investigator for the National

Science Foundation grant supporting the development of environmental ethics for global marine ecosystems. As a member of the University of Georgia Environmental Ethics Certificate Program, Ms. Dallmeyer instructs courses in environmental dispute resolution and marine environmental ethics.

**Richard B. Deriso** is currently an Associate Adjunct Professor of Biological Oceanography at the Scripps Institution of Oceanography and the Chief Scientist of the Tuna-Billfish Program at the Inter-American Tropical Tuna Commission (IATTC). He received his Ph.D. in Biomathematics from the University of Washington. Dr. Deriso's research interests include population dynamics, quantitative ecology, and fishery stock assessment. A former member of the Ocean Studies Board, he has also served as Co-chair for the NRC Committee on Fish Stock Assessment Methods and as a member of two other NRC committees: a Review of Atlantic Bluefin Tuna, and Cooperative Research in the National Marine Fisheries Service.

**James H. Cowan, Jr.**, is a professor in both the Department of Oceanography and Coastal Sciences and the Coastal Fisheries Institute at Louisiana State University. He received an M.S. in Biological Oceanography from Old Dominion University, and both an M.S. in Experimental Statistics and a Ph.D. in Marine Sciences from Louisiana State University. His current research interests include fisheries ecology, biological and fisheries oceanography, biometrics, food-web dynamics, and population demographics and genetics. Dr. Cowan has served as a U.S. delegate both to the International Council for the Exploration of the Sea and the Pacific Marine Sciences Organization. He was Chairman of the Reef Fish Stock Assessment Panel from 1992 to 2004 and is currently a member of the Scientific and Statistical Committee for the Gulf of Mexico Fishery Management Council. Dr. Cowan also previously served on the NRC Committee to Review Individual Fishing Quotas.

**Larry B. Crowder** is Professor of Marine Ecology at the Nicholas School for the Environment at Duke University. He completed his doctoral studies in Zoology at Michigan State University. Dr. Crowder's research centers on predation and food-web interactions, mechanisms underlying recruitment variation in fishes, and population modeling in conservation biology. He has studied food-web processes in estuaries and lakes and has used observational, experimental, and modeling approaches to understand these interactions in an effort to improve fisheries management. Recently Dr. Crowder has begun developing more extensive programs in marine conservation, including research on bycatch, nutrients and low oxygen, marine invasive species, and integrated ecosystem management. Dr. Crowder is a former member of the Ocean Studies Board and has served on the NRC's U.S. National Scientific Committee on Oceanic Research and the Committee on the Alaska Groundfish Fishery and Steller Sea Lions.

**Robert T. Paine** recently retired (1998) from his position as Professor and former Chairman of the Zoology Department at the University of Washington, where he had worked since 1962. Dr. Paine earned his Ph.D. from the University of Michigan in 1961. His research interests include experimental ecology of organisms on rocky shores, interrelationships between species in an ecosystem, and the organization and structure of marine communities. He has examined the roles of predation and disturbance in promoting coexistence and biodiversity. Dr. Paine is a member of the National Academy of Sciences and the Ocean Studies Board and was previously a member of the Board on Life Sciences. He has had extensive experience with the National Research Council; his most recent committee service includes the Committee on Protection of Ecology and Resources of the Caspian Sea and chairing the Committee on the Alaska Groundfish Fishery and Steller Sea Lions.

**Ana M. Parma** is a research scientist with CONICET—the Argentine Council for Science & Technology, working at the Centro Nacional Patagónico in the south of Argentina. She earned her Ph.D. in Fisheries Science in 1989 from the University of Washington and worked as an assessment scientist at the International Pacific Halibut Commission until 2000. Dr. Parma's research interests include fish stock assessment, population dynamics, and adaptive management of fisheries resources. The main focus of her work is on small-scale coastal reef and shellfish fisheries, where she is involved in the evaluation and implementation of spatially explicit management approaches in several fisheries in South America. Among her many prestigious honors, Dr. Parma was awarded a Pew Fellowship in Marine Conservation in 2003 and was appointed Mote Eminent Scholar in 2003. Her previous experience with the Ocean Studies Board includes service on three committees: the Committee on Evaluation, Design, and Monitoring of Marine Reserves and Protected Areas in the United States; the Committee on Fish Stock Assessment Methods; and the Committee to Review Northeast Fishery Stock Assessments.

**Andrew A. Rosenberg** is a Professor of Natural Resources and Earth, Oceans, and Space at the University of New Hampshire. He received his Ph.D. from Dalhousie University in Halifax, Nova Scotia. Dr. Rosenberg explores marine sciences, marine policy, and fisheries in his research projects. Even before joining the University of New Hampshire, he engaged in a distinguished career involving marine sciences. As former Deputy Director of the National Marine Fisheries Service, Dr. Rosenberg was a key policy maker for that agency and served as a liaison to Congress, senior levels of the administration, resource management partners, and the public. Prior to the deputy director post, he served the National Marine Fisheries Service for ten years, where he was the Northeast Regional Administrator and Chief of Research Coordination in Maryland and Massachusetts. Most recently, Dr. Rosenberg served as a member of the President-appointed U.S. Commission on Ocean Policy.

**James E. Wilen** is Director of the Center for Natural Resource Policy Analysis and Professor of Agricultural and Resource Economics at the University of California, Davis. Dr. Wilen received his B.A. degree from California State University and his Ph.D. from the University of California, Riverside. His research focuses broadly on natural resource economics. His specific interests include bioeconomic modeling of fisheries systems, dynamics of exploitation patterns, factor distortion under regulated open access, natural resource damage analysis, and spatial models of fisheries systems. His most recent research focuses on the economics of marine reserves, spatial fisheries management, and comparative analysis of fisheries policies. Dr. Wilen previously served on the Ocean Studies Board Committee on Evaluation, Design, and Monitoring of Marine Reserves and Protected Areas in the United States.

## NATIONAL RESEARCH COUNCIL STAFF

**Christine Blackburn** worked as a program officer with the Ocean Studies Board until early 2006. She earned her Ph.D. in oceanography from the Scripps Institution of Oceanography. Since receiving her doctorate, Dr. Blackburn has been working in science policy, first as a Sea Grant Policy Fellow at the California Resources Agency. She then received an AAAS Science Policy Fellowship which brought her to Washington, DC to work at the National Institutes of Health. In 2003, Dr. Blackburn became a policy associate for the U.S. Commission on Ocean Policy where she helped with the monumental task of preparing the Commission's 500+ page report (and 6 separate volumes of appendixes). She is currently the Ocean Program Project Manager at the California Coastal Conservancy.

**Susan Park** is an associate program officer with the Ocean Studies Board. She earned a Ph.D. in oceanography from the University of Delaware in 2004. Her dissertation focused on the range expansion of the nonnative Asian shore crab *Hemigrapsus sanguineus*. In the summer of 2002, she participated in the Christine Mirzayan Science and Technology Graduate Policy Fellowship with the Ocean Studies Board. During her fellowship, she worked on the OSB study on Nonnative Oysters in the Chesapeake Bay. Prior to joining the Ocean Studies Board in 2006, Dr. Park spent time working on aquatic invasive species management with the Massachusetts Office of Coastal Zone Management and the Northeast Aquatic Nuisance Species Panel.

**Nancy Caputo** is a research associate at the Ocean Studies Board, where she has worked since 2001. Ms. Caputo received an M.P.P. (Master of Public Policy) from the University of Southern California, and a B.A. in political science/international relations from the University of California at Santa Barbara. Her interests include marine policy, science, and education. During her tenure with OSB, Ms. Caputo has assisted with the completion of six reports: *A Review of the*

*Florida Keys Carrying Capacity Study* (2002), *Emulsified Fuels—Risks and Response* (2002), *Decline of the Steller Sea Lion in Alaskan Waters—Untangling Food Webs and Fishing Nets* (2003), *Enabling Ocean Research in the 21st Century: Implementation of a Network of Ocean Observatories* (2003), *River Basins and Coastal Systems Planning Within the U.S. Army Corps of Engineers* (2004), and *Charting the Future of Methane Hydrate Research in United States* (2004). She is also the assistant editor of *Oceanography*, the professional magazine of The Oceanography Society.

**Phillip Long** earned a B.S. in Chemistry and a B.A. in History from the University of Portland. During college, Mr. Long was a student teacher in calculus, physics, and chemistry, and spent one summer working as a research assistant in an inorganic chemistry lab at Texas A&M University. Since graduating in May 2003, he worked as a medical research assistant at the Oregon Health Sciences University and then wandered north to Alaska where he worked in the Moose's Tooth Pub in Anchorage. Adventure, ironically, brought Mr. Long east where he worked as a Program Assistant for the Ocean Studies Board starting in December 2004. Mr. Long commenced another position within the National Academies in February 2006 as a Senior Program Assistant with the Board on Physics and Astronomy.

# Appendix B

# List of Acronyms

| | |
|---|---|
| APB | Attitudes, perceptions, and beliefs |
| CCSBT | Commission for the Conservation of Southern Bluefin Tuna |
| CITES | Convention on International Trade in Endangered Species |
| EPA | Environmental Protection Agency |
| ESA | Endangered Species Act |
| EwE | Ecopath-with-Ecosim |
| FAO | Food and Agriculture Organization (United Nations) |
| FIB | fishing-in-balance |
| $F_{MSY}$ | Fishing Mortality Rate |
| GIS | Geographical Information System |
| IATTC | Inter-American Tropical Tuna Commission |
| IFQ | Individual fishing quota |
| IQ | Individual quota |
| ITQ | Individual transferable quota |
| IUCN | The World Conservation Union (International Union for Conservation of Nature and Natural Resources) |
| IWC | International Whaling Commission |
| LTER | Long Term Ecological Research |

| | |
|---|---|
| MMPA | Marine Mammal Protection Act |
| MPA | Marine Protected Area |
| MSFCMA | Magnuson-Stevens Fishery Conservation and Management Act |
| MSY | Maximum Sustainable Yield |
| | |
| NAO | North Atlantic Oscillation |
| NCEAS | National Center for Ecological Analysis and Synthesis |
| NMFS | National Marine Fisheries Service |
| NOAA | National Oceanic and Atmospheric Administration |
| NRC | National Research Council |
| | |
| PDO | Pacific Decadal Oscillation |

# Appendix C

# Committee Meeting Agendas

**Committee on Ecosystem Effects of Fishing: Phase II—Assessments of the Extent of Ecosystem Change and the Implications for Policy**

**Meeting 1**

**National Academies Facility**
**500 Fifth Street, N.W.**
**Washington, DC 20001**
**March 29–30, 2005**

### AGENDA

**Tuesday, March 29-Keck 109, The National Academies**

*Closed Session*

| | |
|---|---|
| 8:00 a.m. | COMMITTEE BREAKFAST |
| 8:30 a.m. | CLOSED SESSION |
| 12:00 p.m. | COMMITTEE LUNCH |

*Open Session*

| | |
|---|---|
| 1:00 p.m. | Welcome and Introductions—**John Magnuson**, *Chair* and **Christine Blackburn**, *Study Director* |
| 1:15 p.m. | Sponsor Comments<br>• **Michael Sissenwine**, *National Oceanic and Atmospheric Administration (NOAA)*<br>• **Steve Murawski**, *National Oceanic and Atmospheric Administration (NOAA)*<br>Presentation: Ecosystem Effects of Fishing: Phase II—Assessments of the Extent of Ecosystem Change and the Implications for Policy |
| 2:00 p.m. | Discussion with questions and answers |
| 2:45 p.m. | BREAK |
| 3:00 p.m. | **Michael Fogarty**, *Northeast Fisheries Science Center, NOAA*<br>Presentation: Changes in the Structure of the Georges Bank Ecosystem |
| 3:30 p.m. | **James Kitchell**, *University of Wisconsin*<br>Presentation: Lessons from Ecosystem Modeling |
| 4:00 p.m. | **Jeremy Collie**, *University of Rhode Island*<br>Presentation: Regime Shifts and the Recovery of Marine Fish Populations |
| 4:30 p.m. | **Alison Rieser**, *University of Maine*<br>Presentation: Institutional Reforms for Restoring Fisheries-altered Ecosystems |
| 5:00 p.m. | GENERAL DISCUSSION AND QUESTIONS |
| 6:00 p.m. | Open Session adjourns for the day |

APPENDIX C

## Wednesday, March 30-Keck 109, The National Academies

*Closed Session*

8:00 a.m.   Committee Reconvenes

5:00 p.m.   Meeting Adjourns

## Committee on Ecosystem Effects of Fishing: Phase II—Assessments of the Extent of Ecosystem Change and the Implications for Policy

### Meeting 2

The Warwick Seattle Hotel
401 Lenora Street
Seattle, WA 98121
May 9–10, 2005

### AGENDA

## Monday, May 9-Cambridge Conference Room, The Warwick Seattle Hotel

*Closed Session*

8:00 a.m.   Committee Reconvenes

*Open Session*

10:30 a.m.   Welcome and Introductions—**John Magnuson**, *Chair* and **Christine Blackburn**, *Study Director*

10:35 a.m.   **David Fluharty**, *University of Washington*
Presentation: Framing the Policy Question, Generating Scientific Advice and Getting Institutions to Listen

11:05 a.m.   **Janis Searles**, *Oceana*
(Was not able to attend)

12:00 p.m.   LUNCH (Provided for committee, speakers, and staff only)

1:00 p.m.   **Anne Hollowed**, *Alaska Fisheries Science Center*
Presentation: Challenges to Implementing Ecosystem Approaches to Management in the North Pacific

1:45 p.m. **Villy Christensen**, *University of British Columbia*
Presentation: Using Ecosystem Modeling to Evaluate Ecosystem Effects of Fishing: Status and Outlook

2:30 p.m. **Daniel Pauly**, *University of British Columbia*
Presentation: The Marine Trophic Index as Ecosystem Indicator: Implications for Research (and Management?)

3:15 p.m. BREAK

3:30 p.m. **William Sydeman**, *Point Reyes Bird Observatory*
Presentation: Seabirds Indicate Ecosystem Effects of (for) Fishing

4:15 p.m. **Daniel Huppert**, *University of Washington*
(Was not able to attend)

5:00 p.m. Open Session adjourns for the day

**Tuesday, May 10-Cambridge Conference Room, The Warwick Seattle Hotel**

*Closed Session*

8:00 a.m. Committee Reconvenes

5:00 p.m. Meeting Adjourns

**Committee on Ecosystem Effects of Fishing: Phase II—Assessments of the Extent of Ecosystem Change and the Implications for Policy**

**Meeting 3**

**National Academies Facility
500 Fifth Street, N.W.
Washington, DC 20001
June 30–July 1, 2005**

**AGENDA**

**Thursday, June 30-Keck 208, The National Academies**

*Closed Session*

8:00 a.m.– Committee meets in closed session
12:00 p.m.

*Open Session*

12:00 p.m.   LUNCH (provided for committee, speakers, and staff only)

1:00 p.m.   **Tim Essington**, *University of Washington*
Presentation: Patterns and Consequences of "Fishing Down the Food Web": A Comparative Analysis of Fisheries Expansion

1:45 p.m.   **Phil Levin**, *National Oceanic and Atmospheric Administration*
Presentation: EMOCC—an Ecosystem Model of the California Current: Moving from Fisheries Ecology to Ecosystem-based Management

2:30 p.m.   **Josh Eagle**, *University of South Carolina*
Presentation: Ecosystem Effects of Fishing: Policy Implications

3:30 p.m.   Open session adjourns

*Closed Session*

3:30 p.m.–5:00 p.m.   Committee meets in closed session

## Friday, July 1- Keck 109, The National Academies

*Closed Session*

8:00 a.m.–5:00 p.m.   Committee meets in closed session

# Appendix D

# Glossary

**abundance**: Measure of a population size. The quantity of individuals in a population or in a specific area (e.g., fishing grounds) as expressed in number of fish or in biomass. Abundance can be measured in absolute or relative terms.
**adaptive management**: A management plan that acknowledges the uncertainty of a managed system and therefore integrates design, management, and monitoring in order to allow managers to adapt and to learn.
**anadromous**: Fish that spend their adult life in the sea but swim upriver to freshwater spawning grounds in order to reproduce.
**anoxia**: Absence of oxygen relative to atmospheric levels.

**baseline**: A set of reference data or analyses used for comparative purposes; it can be based on a reference year or a reference set of (standard) conditions.
**Bayesian**: A formal statistical approach in which expert knowledge or beliefs are analyzed together with data. Bayesian methods make explicit use of probability for quantifying uncertainty. Bayesian methods are particularly useful for making decision analyses.
**benthic**: The bottom of a waterbody; organisms that live on or in the bottom of a waterbody.
**benthic-pelagic coupling**: The cycling of nutrients between the bottom sediments and the overlying water column.
**biomass**: The total weight of a stock or population of organisms at a given point in time, usually in pounds or metric tons (2,205 pounds = 1 metric ton).
**biotic**: Relating to life and living organisms.

**bottom-up management**: A process of management in which information and decisions are decentralized and in which resource users actively participate in the decision-making process.

**bycatch**: The portion of a fishing catch that is discarded as unwanted or commercially unusable.

**cascading effect (or trophic cascade)**: A food web phenomenon in which changes in abundance at a higher trophic level lead to changes in abundance at lower trophic levels.

**catch**: The total number (or weight) of organisms caught by fishing operations. Catch should include all organisms killed by the act of fishing, not just those landed.

**catch control rule**: A formula used to determine the catch quota as a function of some specific indicators of stock status and any other variable condition used to adjust annual harvest targets. The control rule provides numeric guidance for adjusting catch rates to track forecasts of fluctuations in stock abundance and to achieve management goals. In many fisheries, it is the primary mechanism for regulating harvest rates.

**catch quota**: A limit placed on the total catch allowed within a particular period of time.

**commercial fisheries**: Harvesting fish for profit. This includes those caught for sale, barter, and trade.

**decadal oscillation**: Cyclical changes where shifts occur on scales of roughly 10 years.

**degrade**: To reduce in value or level. In this context, degraded is used to describe ecosystems that have been exploited to a point where there is a loss of desired uses, including a reduction in overall productivity or the loss of species.

**depensatory**: A situation in which mortality rate increases and/or reproduction decreases as the size of the population decreases.

**Ecopath model**: An ecological/ecosystem modeling software used to develop a static, mass-balanced representation of the feeding interactions and nutrient flows in an aquatic ecosystem.

**effort**: The amount of fishing gear of a specific type used on the fishing grounds over a given unit of time; e.g., hours trawled per day, number of hooks set per day or number of hauls of a beach seine per day. When two or more kinds of gear are used, the respective efforts must be adjusted to some standard type before being added.

**El Niño**: Abnormally warm ocean climate conditions which, in some years, affect the Eastern coast of Latin America (centered on Peru) often around Christmas time and which occasionally can be transmitted northward to Alaskan waters. The anomaly is accompanied by dramatic changes in species abundance and distribution, higher local rainfall and flooding, and massive deaths of fish and their predators (including birds). Many other climatic anomalies around the world (e.g., droughts, floods, forest fires) are attributed to consequences of El Niño.

**elasmobranch**: A group of fish without hard bony skeletons, including sharks, skates, and rays.

**eutrophication**: Generally, the natural or man-made process by which a body of water becomes enriched in dissolved mineral nutrients (particularly phosphorus and nitrogen) that stimulate the growth of aquatic plants and enhances organic production of the water body. Excessive enrichment may result in the depletion of dissolved oxygen and eventually to species mortality.

**EwE (Ecopath with Ecosim models)**: Ecological/ecosystem software used to model the dynamics of the feeding interactions and nutrient flows in an exploited aquatic ecosystem. It is used to investigate the structural and functional attributes of each food web and to analyze the effects of alternative harvesting strategies on them.

**exploitation rate**: The proportion of a population whose mortality was caused by fishing, usually expressed in an annual value.

**food web**: The network of feeding relationships within an ecosystem or a community (i.e., the predator-prey relationships) that determines the flow of energy and materials from plants to herbivores, carnivores, and scavengers.

**genotype**: The genetic make-up of an individual, different from its physical appearance (phenotype).

**growth overfishing**: Fishing mortality in which the losses in weight from total mortality exceed the gain in weight due to growth. Growth overfishing results from catching too many small fish before they reached an optimum marketable size.

**harvest**: The total number or poundage of fish caught and kept from an area over a period of time.

**hysteresis**: The lag between making a change and the response to the change.

**impact analysis**: A modeling system that identifies the impact of a change using costs and benefits.

**individual quota**: The share of a total allowable catch (TAC) assigned to an individual, a vessel, or a company. If an individual quota is transferable, it is referred to as an Individual Transferable Quota (ITQ).

**individual transferable quota (ITQ)**: A type of individual quota allocated to individual fishermen or vessel owners and which can be bought and sold once distributed.

**input control**: Management instruments used to control the time and place as well as type and/or amount of fishing to limit yields and fishing mortality; e.g., restrictions on type and quantity of gear, effort, and capacity; closed seasons.

**interaction webs**: A food web diagram depicting the strength of interactions among species; species are linked by arrows indicating the general consequences of altering the abundance or mass of critically important species.

**intertidal assemblage**: A group of co-occurring populations, including both plant and animal species, living between the high and low water levels on marine shores

**iteroparous**: A life history in which individuals reproduce more than once in a lifetime.

**keystone species**: Individual species whose removal may engender dramatic changes in the structure and functioning of a biological community.

**logbook**: A detailed, usually official record of a vessel's fishing activity registered systematically on board the fishing vessel, usually including information on catch and its species composition, the corresponding fishing effort, and location. Completion of logbooks may be a compulsory requirement for a fishing license.

**marine protected area (MPA)**: Geographic area with discrete boundaries that has been designated to enhance the conservation of marine resources. This includes MPA-wide restrictions on some activities such as oil and gas mining and the use of zones such as fishery and ecological reserves to provide higher levels of protection.

**maximum sustainable yield (MSY)**: The largest average catch or yield that can continuously be taken from a stock under existing environmental conditions without significantly affecting the reproduction process. The MSY is also referred to as maximum equilibrium catch, maximum sustained yield, and sustainable catch.

**mean size**: The average size of any particular group of fish.

**megafauna**: Large or relatively large animals, as of a particular region or period, considered as a group.

**mortality**: Measure of death rate.

**North Atlantic Oscillation (NAO)**: A complex climatic phenomenon in the North Atlantic Ocean especially associated with fluctuations of climate between Iceland and the Azores. It is characterized predominantly by cyclical fluctuations of air pressure and changes in storm tracks across the North Atlantic.

APPENDIX D                                                                              151

**optimum yield**: The harvest level for a species that achieves the greatest overall benefits, including economic, social, and biological considerations. Optimum yield is different from maximum sustainable yield in that the MSY considers only the biology of the species. The term includes both commercial and sport yields.

**output control**: Management instruments aimed at directly limiting fish catch or landings through regulation of the total allowable catch and quotas.

**Pacific Decadal Oscillation (PDO)**: A decadal (20-30 year) pattern of Pacific climate variability. The PDO is detected as warm or cool surface waters in the Pacific Ocean, north of 20°N. During a "warm," or "positive," phase, the west Pacific becomes cool and part of the eastern Pacific ocean warms.

**paleoecology**: The branch of ecology that deals with the interaction between ancient organisms and their environment.

**pelagic**: Organisms that spend most of their life within the water column with little contact with or dependency on the bottom. May refer to only certain life stages of a species.

**perturbation**: A physical or biological disturbance to a biological assemblage that can be rcognized by changes in species distributions or abundances.

**phenotype**: The detectable outward manifestation of a specific genetic trait or genotype.

**piscivore**: An organism that eats mainly fish.

**planktivore**: An organism that consumes plankton.

**population**: Organisms of the same species that occur in a particular place at a given time. A population may contain several discrete breeding groups or stocks.

**pristine**: An environmental state in which anthropogenic influences are thought to be non-existent.

**purse seine**: A fishing net with a line at the bottom that enables the net to be closed like a purse. Purse seines are very large and can be used to catch entire schools of fish.

**real-time data**: Data which are reported almost simultaneously with collection. Real-time data are the most current information available, being collected and posted at essentially the same moment.

**recreational fisheries**: The harvesting of fish for personal use, fun, and challenge (i.e., as opposed to harvest for profit or research). Recreational fishing does not include sale, barter, or trade of all or part of the catch.

**recruitment**: A measure of the number of fish that enter a class during some time period, such as the spawning class or fishing-size class.

**regime shifts**: A medium- or long-term shift in environmental conditions that impacts the productivity of a stock.

**rehabilitation**: To improve the quality of a habitat or ecosystem, but not necessarily to fully restore all functions to their undisturbed condition.

**restoration**: To return an ecosystem to a close approximation of its condition prior to disturbance. The goal is to emulate a natural, functioning, self-regulating system that is integrated with the ecological landscape in which it occurs.

**sectoral management**: Management approach in which specific agencies are given responsibility for managing particular sectors (e.g., fisheries, tourism, water quality), as opposed to integrated management in which various sectors are considered together. The result of sectoral management of an area in which different sectors compete for resources is often conflict between users, and between different sector management agencies with responsibilities over a common area, even under the same government. There is an inherent incentive for each sector to maximize its profits and benefits at the expense of other sectors, the general public, or the natural environment.

**sequential (or serial) addition**: A term referring to the addition of new species to a fishery when stocks of the previous fishery species become depleted. The sequential addition of lower-trophic level species along with upper-trophic-level fisheries within an ecosystem is also known as "fishing through a food web."

**sequential (or serial) depletion**: A term referring to the systematic loss of species in a commercial fishery due to overfishing. The serial depletion of higher-trophic-level fisheries, and subsequent replacement with lower-trophic level species, is also known as "fishing down the food web."

**Southern Oscillation**: An oscillation in air pressure between the southeastern and southwestern Pacific waters. When the eastern Pacific waters increase in temperature (an El Niño event), atmospheric pressure rises in the western Pacific and drops in the east. This pressure drop is accompanied by a weakening of the easterly Trade Winds. Together with El Niño, this phenomenon is known as ENSO, or El Niño-Southern Oscillation.

**species density**: The number of individuals per unit area or volume.

**stable state**: A property of a community that, if the community is disturbed, it will tend to revert back to its original equilibrium state. It is possible for communities to have more than one stable state and different disturbances will drive community compositions to different alternative stable states.

**stock assessment**: The process of collecting and analyzing biological and statistical information to determine the changes in the abundance of fishery stocks in response to fishing, and, to the extent possible, to predict future trends of stock abundance. Stock assessments are based on resource surveys; knowledge of the habitat requirements, life history, and behavior of the species; the use of environmental indices to determine impacts on stocks; and catch statistics. Stock assessments are used as a basis to assess and specify the present and probable future condition of a fishery.

**sustainability**: Characteristic of resources that are managed so that the natural capital stock is non-declining through time while production opportunities are maintained for the future.

**sustainable yield**: The number or weight of fish in a stock that can be taken by fishing without reducing the stock biomass from year to year, assuming that environmental conditions remain the same.

**telemetry**: The collection and transmission of data from remote locations to a central station.

**Thunnids**: Tuna species.

**time-series**: Measurements of data over time arranged in order of occurrence. Time series are often used to project future values by observing how the value of a variable has changed in the past.

**top-down management**: A process of management in which information and decisions are centralized and in which resource users are kept outside of the decision-making process.

**total allowable catch (TAC)**: The annual recommended catch for a species or species group. The regional council sets the TAV from the range of allowable biological catch.

**total marine capture fisheries production**: The total global harvest of marine capture fisheries (capture fisheries do not include production from aquaculture).

**trophic level**: Position in food chain determined by the number of energy-transfer steps to that level. Plant producers constitute the lowest level, followed by herbivores and a series of carnivores at the higher levels.

**trophospecies**: A group of species with similar trophic roles; i.e., species with similar foods and predators.

**utilized stock**: The number of individuals within a stock that are alive at a given time but which will be caught in the future.

**year class**: Individuals in a population that were born in the same year. For example, the 1987 year class of cod includes all cod born in 1987, which would be age 1 in 1988. Occasionally, a stock produces a very small or very large year class which can be pivotal in determining stock abundance in later years.

**yield**: The production from a fishery, often given in weight. Catch and yield are often used interchangeably.